METHODS IN MOLECULAR BIOLOGY™

Series Editor
John M. Walker
School of Life Sciences
University of Hertfordshire
Hatfield, Hertfordshire, AL10 9AB, UK

For other titles published in this series, go to
www.springer.com/series/7651

Cyclooxygenases

Methods and Protocols

Edited by

Samir S. Ayoub, Roderick J. Flower, and Michael P. Seed

Centre for Biochemical Pharmacology, Experimental Pathology Group,
St. Bart's and the London School of Medicine and Dentistry, William Harvey Research Institute,
Queen Mary University of London, London, UK

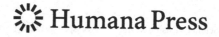

Humana Press

Editors
Samir S. Ayoub, Ph.D.
Centre for Biochemical Pharmacology,
  Experimental Pathology Group
St. Bart's and the London School of Medicine
  and Dentistry
William Harvey Research Institute
Queen Mary University of London
London, UK
s.s.ayoub@qmul.ac.uk

Roderick J. Flower, Ph.D.
Centre for Biochemical Pharmacology
St. Bart's and the London School of Medicine
  and Dentistry
William Harvey Research Institute
Queen Mary University of London
London, UK
r.j.flower@qmul.ac.uk

Michael P. Seed, Ph.D.
Centre for Biochemical Pharmacology,
  Experimental Pathology Group
St. Bart's and the London School of Medicine
  and Dentistry
William Harvey Research Institute
Queen Mary University of London
London, UK
m.p.seed@qmul.ac.uk

ISSN 1064-3745
ISBN 978-1-4939-5638-8
DOI 10.1007/978-1-59745-364-6
Springer New York Dordrecht Heidelberg London

e-ISSN 1940-6029
ISBN 978-1-59745-364-6 (eBook)

Printed on acid-free paper

Humana Press is a part of Springer Science+Business Media (www.springer.com)

Preface

The nonsteroidal anti-inflammatory drugs (NSAIDs) consist of more than 40 compounds, which are the most widely used group of medicines worldwide for the treatment of inflammation, pain, fever, and for the prevention of thrombosis. It was through the pivotal work of the late Professor Sir John Vane in the early 1970s that it was realized that the mechanism of pharmacological and toxicological actions of these compounds depended on the inhibition of the catalytic activity of the cyclooxygenase (COX) enzyme resulting in the reduction of prostaglandin synthesis. After the discovery of the second isoform of COX, named COX-2 in the early 1990s, it became clear that the pharmacological actions of NSAIDs were due to the inhibition of COX-2, while the toxicological actions were due to inhibition of COX-1. Thus COX-1 became known as the "house-keeping" isoform involved in prostaglandin synthesis during physiological processes such as regulation of acid release in the stomach, while COX-2 as the inducible isoform involved in pathological processes such as inflammation, pain, fever, and cancer. Eventually, this led to the development of the COX-2 selective inhibitors, of which celecoxib was the first.

Advancement in COX research and their bioactive products, the prostanoids, was made possible by the development of new techniques and improvement of existing techniques specific to COX or general techniques applied to COX research. These techniques include the purification (Chapter 3), cloning (Chapter 4), and expression (Chapters 4 and 5) of COX enzymes as well as determining the activity of COX enzymes in different experimental conditions (Chapters 6, 7, 9–11). COX activity assay techniques, especially cell based ones (Chapter 8), along with measurement of prostaglandins (Chapter 13) have been used to determine the relative potencies of NSAIDs on the inhibition of COX-1 and COX-2. In vivo models of gastrointestinal injury (Chapter 16) and of inflammation (Chapter 17) and pain (Chapter 18) have been applied to COX research and have helped in elucidating the role played by COX-1 and COX-2 in these disease processes.

Cyclooxygenases: Methods and Protocols is a collection of in vitro and in vivo techniques routinely used in COX research. This book is a laboratory manual aimed at scientists interested in applying these techniques to their research programs. *Cyclooxygenases: Methods and Protocols* is written by some of the scientists who have pioneered and have helped to shape the COX field. It is our hope, as editors, that you will find this book a useful tool.

Finally, the editors would like to thank all the contributors and in particular Professor John Walker, as Series Editor, for inviting us to edit this book and for his help in reviewing manuscripts and putting this book together.

London, UK

Samir S. Ayoub, Ph.D.
Roderick J. Flower, Ph.D., FRS
Michael P. Seed, Ph.D.

Contents

Contributors

SAMIR S. AYOUB • *Centre for Biochemical Pharmacology, Experimental Pathology Group, St. Bart's and the London School of Medicine and Dentistry, William Harvey Research Institute, Queen Mary University of London, London, UK*

REGINA M. BOTTING • *St. Bart's and the London School of Medicine and Dentistry, William Harvey Research Institute, Queen Mary University of London, London, UK*

OLIVIER BOUTAUD • *Department of Pharmacology, Vanderbilt University, Nashville, TN, USA*

MARTA L. CAPONE • *Department of Medicine and Center of Excellence on Aging, G. d'Annunzio University School of Medicine, and G. d'Annunzio University Foundation, Chieti, Italy*

KELSEY C. DUGGAN • *A.B. Hancock Jr. Memorial Laboratory for Cancer Research, Departments of Biochemistry, Chemistry, and Pharmacology, Center in Molecular Toxicology, Vanderbilt-Ingram Cancer Center, Vanderbilt Institute of Chemical Biology, Vanderbilt University School of Medicine, Nashville, TN, USA*

RODERICK J. FLOWER • *Centre for Biochemical Pharmacology, St. Bart's and the London School of Medicine and Dentistry, William Harvey Research Institute, Queen Mary University of London, London, UK*

JAMES K. GIERSE • *Inflammation Research, Pfizer Global Research and Development, Chesterfield, MO, USA*

DEREK W. GILROY • *Division of Medicine, Centre for Clinical Pharmacology, Rayne Institute, University College London, London, UK*

WILLIAM F. HOOD • *Pfizer Global Research and Development, Chesterfield, MO, USA*

JOHN C. HUNTER • *Department of Chemistry and Biochemistry, Brigham Young University, Provo, UT, USA*

HIROYASU INOUE • *Department of Food Science and Nutrition, Faculty of Human Life & Environment, Nara Women's University, Kita-Uoya-Nishimachi, Nara, Japan*

STEFAN LAUFER • *Pharmazeutisches Institut, Eberhard Karls Universität Tübingen, Tübingen, Germany*

SABINE LUIK • *Pharmazeutisches Institut, Eberhard Karls Universität Tübingen, Tübingen, Germany*

LAWRENCE J. MARNETT • *A.B. Hancock Jr. Memorial Laboratory for Cancer Research, Departments of Biochemistry, Chemistry, and Pharmacology, Center in Molecular Toxicology, Vanderbilt-Ingram Cancer Center, Vanderbilt Institute of Chemical Biology, Vanderbilt University School of Medicine, Nashville, TN, USA*

ADRIAN R. MOORE • *Celltech R&D Limited, Slough, Berkshire, UK*

JOEL MUSEE • *A.B. Hancock Jr. Memorial Laboratory for Cancer Research, Departments of Biochemistry, Chemistry, and Pharmacology, Center in Molecular Toxicology, Vanderbilt-Ingram Cancer Center, Vanderbilt Institute of Chemical Biology, Vanderbilt University School of Medicine, Nashville, TN, USA*

NATALIE M. MYRES • *Department of Chemistry and Biochemistry, Brigham Young University, Provo, UT, USA*

RIEKO NAKATA • *Department of Food Science and Nutrition, Faculty of Human Life & Environment, Nara Women's University, Kita-Uoya-Nishimachi, Nara, Japan*

JUSTINE NEWSON • *Division of Medicine, Centre for Clinical Pharmacology, Rayne Institute, University College London, London, UK*

JOHN A. OATES • *Departments of Pharmacology and Medicine, Vanderbilt University, Nashville, TN, USA*

PAOLA PATRIGNANI • *Sezione di Farmacologia, Dipartimento di Medicina e Scienze dell'Invecchiamento, Università di Chieti "G. d'Annunzio", Chieti, Italy*

LORENZO POLENZANI • *R&D Pipeline Strategy Manager, Angelini Research Center, Santa Palomba-Pomezia, Rome, Italy*

BEVERLY A. REITZ • *Pfizer, Inc., Chesterfield, MO, USA*

MICHAEL P. SEED • *Centre for Experimental Medicine and Rheumatology, Bart's and the London School of Medicine and Dentistry, William Harvey Research Institute, Queen Mary University of London, London, UK*

DANIEL L. SIMMONS • *Department of Chemistry and Biochemistry, Brigham Young University, Provo, UT, USA*

MELANIE STABLES • *Division of Medicine, Centre for Clinical Pharmacology, Rayne Institute, University College London, London, UK*

NICHOLAS R. STATEN • *Pfizer, Inc., Chesterfield, MO, USA*

STEFANIA TACCONELLI • *Department of Medicine and Center of Excellence on Aging, G. d'Annunzio University School of Medicine, and G. d'Annunzio University Foundation, Chieti, Italy*

MARK C. WALKER • *Pfizer Global Research and Development, Chesterfield, MO, USA*

BRENDAN J.R. WHITTLE • *St. Bart's and the London School of Medicine and Dentistry, William Harvey Research Institute, Queen Mary University of London, London, UK*

Part I

In Vitro Protocols for Studying Expression
and Activity of Cyclooxygenase Enzymes
and Characterisation of Their Inhibition Patterns

In Vitro Cyclo-oxygenase Expression and Activity Protocols: Introduction to Part I

Roderick J. Flower

Abstract

While the prostaglandin field may be said to have originated in the 1930s, it still remains vibrant today. Probably the main driver for this has been the growing realisation of the pathophysiological importance of the eicosanoid family, especially of their role in inflammatory and cardiovascular disease, as well as an understanding of the pharmacology of the drugs used to treat these conditions. With this has come an interest in the techniques and methodology required to study the enzymatic synthesis and metabolism of these fascinating lipids, such that this can be applied to solve the outstanding problems in the field. This introductory chapter provides a brief overview of the development of the cyclo-oxygenase field together with some introductory comments on the content of the book.

Key words: Prostaglandin, Eicosanoids, Cyclo-oxygenase, Cox-1, Cox-2, NSAID, Inflammation, Analytical protocols

1. Discovery and Biosynthesis

It is 75 years since von Euler and others observed that human seminal plasma contained powerful vasodepressor and smooth muscle contractile factors (1). These unexpected findings marked the beginning of a fascinating quest into the biology of those mysterious substances that later became known as the "prostaglandins". It is a quest that is still actively pursued today by academic researchers as well as by the pharmaceutical industry.

Looking back on the history of the field, several obvious landmarks stand out. Of course, there was the elucidation of the original structure of a "prostaglandin" by Bergstrom and SjÖvall in 1957 (2) and the identification – and naming – of prostaglandins E and F in extracts from sheep vesicular glands in 1960 (3, 4).

Samir S. Ayoub et al. (eds.), *Cyclooxygenases: Methods and Protocols*, Methods in Molecular Biology, vol. 644, DOI 10.1007/978-1-59745-364-6_1, © Springer Science+Business Media, LLC 2010

The recognition that the prostaglandins were derived from essential fatty acids was another breakthrough discovery. The demonstration that an enzyme within cells converted substrates such as arachidonic acid into prostaglandins was made simultaneously in 1964 by groups at the Unilever company in the Netherlands and by Bergstrom's Karolinska Institute group in Stockholm (5, 6). These seminal experiments laid the foundation for the study of what became known initially as *prostaglandin synthetase* but which was subsequently renamed the *fatty acid cyclo-oxygenase* enzyme.

Further work, again primarily by the Karolinska and Unilever laboratories, confirmed in 1973 that the reaction which converted essential fatty acids into prostaglandins proceeded via the formation of highly unstable intermediate endoperoxides, subsequently named prostaglandin G_2 and prostaglandin H_2 (7, 8). Against this background came the appreciation of the widespread, almost ubiquitous distribution of prostaglandins and the discovery of some of their more striking biological properties. It was a study of the enzymatic transformation of the endoperoxide intermediates in various tissues that led to the further discovery of labile prostaglandin-like substances, which had astonishing pathphysiological properties, such as thromboxane A_2 (9) and, later, prostacyclin (10). Other discoveries have followed.

2. Cyclo-Oxygenase Biochemistry

However, our main concern here is with the cyclo-oxygenase enzyme itself. It had been recognised since the 1960s that tissue homogenates generated prostaglandins and that the enzyme activity seemed to be located in the "high-speed" membrane fraction, but it wasn't until the mid 1970s that convincing data were obtained about the intracellular location of the cyclo-oxygenase. Experiments such as those performed by Bohman and Larsson in 1975 (11) on the rabbit kidney suggested that the enzyme was associated with the endoplasmic reticulum. This membrane-bound location of the cyclo-oxygenase complex frustrated many early efforts to characterise this haem-containing enzyme. The subsequent discovery that it could be solubilized using detergents was important in providing a more convenient source of enzyme that could be subjected to further analysis. Useful data were gathered by Roth et al. (12) who ingeniously adapted Vane's finding (see below) to assess the molecular weight of the enzyme by using radioactive aspirin, labelled with tritium on the labile acetyl group. After treatment, a tissue extract contained a labelled protein with a molecular weight of 85 kDa.

As a result of the "molecular revolution" we have all benefited enormously from the cloning and sequencing of cyclo-oxygenase in the late 1980s (13, 14) and the extraordinary discovery of a further Cox isoforms in 1990 – again as the result of work by several laboratories published almost simultaneously. Although presaged to some extent by pharmacological data, the actual discovery of the new isoform came from researches into the nature of inducible genes. While investigating the expression of early-response genes in fibroblasts transformed with Rous sarcoma virus, Simmons and his colleagues (15) identified a novel mRNA transcript that coded for a protein that was not identical but which had a high sequence similarity to the seminal vesicle Cox enzyme and suggested that a Cox isozyme had been discovered. Likewise, while studying phorbol-ester-induced genes in Swiss 3T3 cells Herschman and colleagues (16) also observed a novel cDNA species encoding a protein with a predicted structure similar to Cox-1. The same laboratory confirmed that this gene product was indeed a novel Cox and that its induction was inhibited by dexamethasone. Similarly findings were subsequently reported in several other cell types.

The two cyclo-oxygenases were accordingly renamed as Cox-1, referring to the original enzyme isolated from seminal vesicles and subsequently found to be distributed almost ubiquitously and Cox-2, denoting the "inducible" form of the enzyme (although it was expressed basally in the brain and elsewhere). The two genes had a different chromosomal organization in rodents and humans, and Cox-1/Cox-2 mRNA was differentially expressed in human tissues (17). Promoter analysis confirmed a fundamental difference between the two isozymes, with the Cox-2 promoter containing elements strongly reminiscent of those genes that are activated during cellular stress and down-regulated by glucocorticoids, whereas the Cox-1 promoter had the appearance of a "housekeeping" gene (18). Histological and other studies confirmed this apparent division of labour between the two enzymes, and Cox-1 seemed to be the predominant iso-form in healthy gastrointestinal tissue from several species.

Mechanistically, both Cox isoforms are bifunctional enzymes and have two distinct catalytic activities. The first, dioxygenase step incorporates two molecules of oxygen into the arachidonic (or other fatty acid substrate) chain at C-11 and C-15. This allows a cyclisation and the formation of the highly unstable endoperox-ide intermediate prostaglandin G_2, which bears a hydroperoxy group at C-15. The second, peroxidase, function of the enzyme converts this to prostaglandin H_2 having a hydroxygroup at C-15. This moiety can then be transformed in a cell-specific manner by separate isomerase, reductase or synthase enzymes into other prostanoids. Both Cox-1 and Cox-2 probably exist as homodimers attached to intracellular membranes.

The subsequent solution of the crystal structure of any enzyme is itself a landmark achievement, and the case of the cyclo-oxygenases was to shed much light on the mechanism of the selective Cox-2 inhibitors (see below) (19, 20). Despite their high homology, detailed examination of the structure of the catalytic sites revealed the substrate binding "channel" in the two enzymes to be quite different. A single amino-acid change, from the comparatively bulky Ile in Cox-1 to Val at position 523 in Cox-2 (equivalent to position 509 in Cox-1), and the conformational changes that this produced resulted in enhanced access to a bulging "side pocket" in Cox-2 that was not present in Cox-1.

More recently still, further potential cyclo-oxygenase variants (21) have been identified. These may prove important in understanding the biology of the enzyme in the future.

3. Prostaglandins and Aspirin

Interspersed among this work on prostaglandin biosynthesis was another landmark discovery: the demonstration, by Vane (22) and his colleagues, that this enzyme was the target for the aspirin-like drugs (non-steroidal anti-inflammatory drugs; NSAIDs). At a stroke, this transformed our understanding of the pharmacology of these compounds but also provided prostaglandin researchers with a useful set of tools to explore further the biology of the prostaglandins. It also threw into sharp focus the importance of understanding the enzymology of the cyclo-oxygenase.

The late 1960s and early 1970s had seen an explosion of interest in prostaglandin biology. It was observed that prostaglandins were generated during platelet aggregation and that they produced fever, hyperalgesia, and inflammation. Prostaglandins had also been detected in the gastric mucosa and been shown to inhibit ulcer formation in rodent models of gastric damage and cytoprotection. In other words, the ability of NSAIDs to block Cox provided the much sought after conceptual link between the therapeutic and side effects of these drugs (reviewed in (23)).

Over the next couple of years, it was shown that the entire gamut of NSAIDs inhibited Cox at concentrations well within their therapeutic plasma range and that the overall order of potency corresponded with their therapeutic activity. Other types of anti-inflammatories, such as the glucocorticoids and the so-called disease modifying drugs such as gold and penicillamine, were inactive in these cell-free assays, providing further evidence for the specificity of the NSAID effect. Using active/inactive enantiomeric pairs of NSAIDs such as naproxen, an exquisite correlation was observed between the anti-inflammatory and anti-Cox activity, and many more studies confirmed the notion that this is a fundamental mechanism of this class of drugs.

By 1974, Vane's concept was firmly established as the most powerful explanatory idea in the NSAID field. One effect of this was that the pharmaceutical industry now possessed, probably for the first time, a simple and robust in vitro to screen for putative anti-inflammatory compounds. Probably as a result of this, the number of chemical abstracts dealing with potential inhibitors of the Cox enzyme rose markedly, with more than 2,500 per year recorded within a decade.

Most NSAIDs inhibit only the initial dioxygenation reaction, and the structural information obtained from the crystallography projects proved crucial in understanding their detailed mechanism of action. To block these enzymes, NSAIDs must enter the hydrophobic channel forming hydrogen bonds with an Arg residue at position 120, thus preventing substrate fatty acids from entering into the catalytic domain. Aspirin is, however, an anomaly. It enters the active site and acetylates a serine at position 530 irreversibly inactivating Cox-1. This is the basis for aspirin's long lasting effects on platelets (24).

The bulging "side pocket" of Cox-2 allowed the binding of selective inhibitors by providing a docking site for side chains such as the phenylsulphonamide residue of the Searle Monsanto selective inhibitor, SC558. In an elegant demonstration of the crucial nature of this Ile/Val substitution, Gierse et al. (25) showed that mutation of this residue in Cox-2 back to Ile largely prevented selective inhibitors from working.

Such structural data also helped explain differences in the inhibitory kinetics of Cox-1 and Cox-2. There are, as stated above, several distinct mechanisms by which Cox-1 inhibitors can block catalytic activity, but many are of the competitive-reversible type. By contrast, the Cox-2 inhibitors are irreversible, time-dependent inhibitors, partly as a result of the binding of the sulphonamide (or related) moiety into the enzyme "side pocket".

4. The Cox-1/Cox-2 Concept

Most of the biological data suggested that Cox-2 was the predominant inflammatory species, and by implication, the best target for NSAIDs. So was Cox-2 inhibition the true therapeutic modality of NSAIDs? If so, Cox-1 inhibition might account for the side effects, such as gastric irritation and depression of platelet aggregation. This was the notion advanced by more than one group (26, 27) which came to be known as the "Cox-2-bad: Cox-1-good" hypothesis. If true, then the obvious corollary was that a selective Cox-2 inhibitor should be an ideal drug, possessing anti-inflammatory actions but lacking gastric and other side effects.

Most "traditional" NSAIDs were found to be inhibitors of both enzymes, though they varied in the degree to which they inhibited each isoform (28). While medicinal chemistry programmes aiming to identify selective inhibitors were being developed, the field continued to produce data that were, on the whole, supportive to the Cox-1/Cox-2 concept. A number of useful, selective Cox-2 inhibitors were soon described in the literature (reviewed in (29)). SC558 (a celecoxib prototype) showed good efficacy in rodent models of inflammation, fever and pain, whereas the closely related SC560, a selective Cox-1 inhibitor, was ineffective. Celecoxib itself reversed carrageenan-induced hyperalgesia and local prostaglandin production in rats, and a related compound was active intrathecally.

The culmination of this, and many other similar studies, was the launch, in the 1990s, of celecoxib, rofecoxib, and subsequently other selective Cox-2 inhibitors. The therapeutic history of this field has not been entirely happy, but there is no doubt that it remains an influential concept. It is generally accepted that the anti-inflammatory action (and probably most analgesic actions) of the NSAIDs is related to their inhibition of Cox-2 while their unwanted effects – particularly those affecting the gastrointestinal tract – are largely a result of their inhibition of Cox-1. It was an idea, then, of great explanatory force if not as much of outstanding clinical benefit as was originally hoped.

5. The Field Today

Today, it is possible to purchase recombinant Cox enzymes and cyclo-oxygenase null transgenic animals from commercial suppliers and perform assays in vitro and in vivo in a way which most of us, working during the early days of prostaglandin research, could only have dreamt about. Interestingly, however, many of the problems that beset the field then (e.g., the nature of the co-factors involved in cyclo-oxygenase action in vitro, the importance of the assay conditions and so on) continue to be an issue today, particularly when attempting to standardise the results of experiments across laboratories.

In this book, many of these questions are directly addressed in chapters authored by some of the key researchers in the field. There are chapters dealing specifically with the purification cloning and expression of Cox-1 and Cox-2 (Chapters 2–5). We may also read of several techniques to identify inhibitors of the enzyme both using purified enzyme preparations as well as tissue homogenates as well as a review of kinetic methods for studying the enzyme and characterising the binding site (Chapters 8–11). Special considerations, such as the generation of hydroperoxides by Cox and importance of "peroxide tone" in regulating catalytic activity, are covered in Chapters 6 and 7.

There are also chapters explaining the best methods for extracting and measuring the products of the cyclo-oxygenase, the eicosanoids themselves from a variety of tissues (Chapters 12–14) and, in the third part of the book, the use of in vivo models to study the involvement of cyclo-oxygenase products in health and disease (Chapters 16–18) is discussed.

An understanding of the cyclo-oxygenase enzyme and the reactions it catalyzes has provided the keys that unlock the mystery of NSAID action as well as aiding the design of more potent and selective drugs. In this respect, it is astonishing to reflect that such a simple drug as aspirin can have such a dramatic action on a complex pathological response such as inflammation, which is so heavily driven by redundant arrays of mediators.

The editors hope that by setting forth in this volume the appropriate technical strategies they will enable the reader to set up their own prostaglandin laboratory and to continue to research this fascinating and productive area.

References

1. Euler USV (1934) Information on the pharmacological effect of natural secretions and extracts from male accessory glands. Arch Exp Pathol Pharm 175:78–84
2. Bergstrom S, Sjovall J (1957) The isolation of prostaglandin. Acta Chem Scand 11:1086
3. Bergstrom S, Sjovall J (1960) The isolation of prostaglandin E from sheep prostate glands. Acta Chem Scand 14:1701–1705
4. Bergstrom S, Sjovall J (1960) The isolation of prostaglandin F from sheep prostate glands. Acta Chem Scand 14:1693–1700
5. Van D, Beerthuis RK, Nugteren DH, Vonkeman H (1964) The biosynthesis of prostaglandins. Biochim Biophys Acta 90:204–207
6. Bergstrom S, Danielsson H, Samuelsson B (1964) The enzymatic formation of prostaglandin E2 from arachidonic acid: prostaglandins and related factors 32. Biochim Biophys Acta 90:207–210
7. Nugteren DH, Hazelhof E (1973) Isolation and properties of intermediates in prostaglandin biosynthesis. Biochim Biophys Acta 326:448–461
8. Hamberg M, Samuelsson B (1973) Detection and isolation of an endoperoxide intermediate in prostaglandin biosynthesis. Proc Natl Acad Sci USA 70:899–903
9. Hamberg M, Svensson J, Samuelsson B (1975) Thromboxanes: a new group of biologically active compounds derived from prostaglandin endoperoxides. Proc Natl Acad Sci USA 72:2994–2998
10. Moncada S, Gryglewski R, Bunting S, Vane JR (1976) An enzyme isolated from arteries transforms prostaglandin endoperoxides to an unstable substance that inhibits platelet aggregation. Nature 263:663–665
11. Bohman SO, Larsson C (1975) Prostaglandin synthesis in membrane fractions from the rabbit renal medulla. Acta Physiol Scand 94:244–258
12. Roth GJ, Stanford N, Majerus PW (1975) Acetylation of prostaglandin synthase by aspirin. Proc Natl Acad Sci USA 72:3073–3076
13. DeWitt DL, el-Harith EA, Smith WL (1989) Molecular cloning of prostaglandin G/H synthase. Adv Prostaglandin Thromboxane Leukot Res 19:454–457
14. Merlie JP, Fagan D, Mudd J, Needleman P (1988) Isolation and characterization of the complementary DNA for sheep seminal vesicle prostaglandin endoperoxide synthase (cyclooxygenase). J Biol Chem 263:3550–3553
15. Xie WL, Chipman JG, Robertson DL, Erikson RL, Simmons DL (1991) Expression of a mitogen-responsive gene encoding prostaglandin synthase is regulated by mRNA splicing. Proc Natl Acad Sci USA 88:2692–2696
16. Kujubu DA, Fletcher BS, Varnum BC, Lim RW, Herschman HR (1991) TIS10, a phorbol ester tumor promoter-inducible mRNA from Swiss 3T3 cells, encodes a novel prostaglandin synthase/cyclooxygenase homologue. J Biol Chem 266:12866–12872
17. O'Neill GP, Ford-Hutchinson AW (1993) Expression of mRNA for cyclooxygenase-1

and cyclooxygenase-2 in human tissues. FEBS Lett 330:156–160

18. Evett GE, Xie W, Chipman JG, Robertson DL, Simmons DL (1993) Prostaglandin G/H synthase isoenzyme 2 expression in fibroblasts: regulation by dexamethasone, mitogens, and oncogenes. Arch Biochem Biophys 306: 169–177

19. Picot D, Loll PJ, Garavito RM (1994) The X-ray crystal structure of the membrane protein prostaglandin H2 synthase-1. Nature 367: 243–249

20. Kurumbail RG, Stevens AM, Gierse JK, McDonald JJ, Stegeman RA, Pak JY et al (1996) Structural basis for selective inhibition of cyclooxygenase-2 by anti-inflammatory agents. Nature 384:644–648

21. Chandrasekharan NV, Dai H, Roos KL, Evanson NK, Tomsik J, Elton TS et al (2002) COX-3, a cyclooxygenase-1 variant inhibited by acetaminophen and other analgesic/antipyretic drugs: cloning, structure, and expression. Proc Natl Acad Sci USA 99:13926–13931

22. Vane JR (1971) Inhibition of prostaglandin synthesis as a mechanism of action for aspirin-like drugs. Nat New Biol 231:232–235

23. Flower RJ (1974) Drugs which inhibit prostaglandin biosynthesis. Pharmacol Rev 26:33–67

24. Roth GJ, Siok CJ (1978) Acetylation of the NH2-terminal serine of prostaglandin synthetase by aspirin. J Biol Chem 253:3782–3784

25. Gierse JK, McDonald JJ, Hauser SD, Rangwala SH, Koboldt CM, Seibert K (1996) A single amino acid difference between cyclooxygenase-1 (COX-1) and -2 (COX-2) reverses the selectivity of COX-2 specific inhibitors. J Biol Chem 271:15810–15814

26. DeWitt DL, Meade EA, Smith WL (1993) PGH synthase isoenzyme selectivity: the potential for safer nonsteroidal antiinflammatory drugs. Am J Med 95:40S–44S

27. Mitchell JA, Akarasereenont P, Thiemermann C, Flower RJ, Vane JR (1993) Selectivity of nonsteroidal antiinflammatory drugs as inhibitors of constitutive and inducible cyclooxygenase. Proc Natl Acad Sci USA 90:11693–11697

28. Warner TD, Giuliano F, Vojnovic I, Bukasa A, Mitchell JA, Vane JR (1999) Nonsteroid drug selectivities for cyclo-oxygenase-1 rather than cyclo-oxygenase-2 are associated with human gastrointestinal toxicity: a full in vitro analysis. Proc Natl Acad Sci USA 96:7563–7568

29. Flower RJ (2003) The development of COX2 inhibitors. Nat Rev Drug Discov 2:179–191

Chapter 2

Techniques Used to Study Regulation of Cyclooxygenase-2 Promoter Sites

Hiroyasu Inoue and Rieko Nakata

Abstract

Cyclooxygenase-2 (COX-2), the rate-limiting enzyme for prostaglandin (PG) biosynthesis, plays a key role in inflammation, tumorigenesis, development and circulatory homeostasis. COX-2 expression is rapidly and sensitively regulated by various stimuli, and also its regulation is distinct among cell types at transcriptional and posttranscriptional levels. Therefore, it is important to consider these features of COX-2 expression in the reporter assays we describe in this chapter. Emphasis should be made with regard to two points. Firstly, COX-2 reporter assays should be evaluated by intrinsic COX-2 expression, such as RT-PCR, northern blotting, western blotting, or by PGE_2 measurement. Secondly, one must carefully choose several conditions in the reporter assays for experimental purposes.

Key words: Cyclooxygenase, Luciferase, Reporter, CRE, NF-IL6, NF-κB, *cis*-Acting element

1. Introduction

Cyclooxygenase (COX) has two isoforms, COX-1 and -2. COX-1 is constitutively expressed in most cells, whereas COX-2 is largely absent but induced upon stimulation by inflammatory stimuli such as endotoxin, lipopolysaccharide (LPS), suggesting that COX-2 plays a critical role in inflammation. Growing evidence indicates that expression of COX-2 is differently regulated in different types of cells ((1–3), Fig. 1).

In COX-2 promoter sites, three *cis*-acting elements, NF-κB, NF-IL6 (also called as C/EBPβ) sites, and cyclic AMP response element (CRE), as well as TATA box are conserved among species (Fig. 2). The CRE is overlapped with E-box with which USF1 and USF2 are known to bind. The CRE in the COX-2 gene is not a symmetrical CRE such as the one found in the somatostatin gene, but an asymmetrical CRE found as a consensus sequence

Samir S. Ayoub et al. (eds.), *Cyclooxygenases: Methods and Protocols*, Methods in Molecular Biology, vol. 644,
DOI 10.1007/978-1-59745-364-6_2, © Springer Science+Business Media, LLC 2010

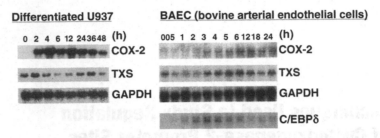

Fig. 1. Distinct expression of COX-2 mRNA by LPS between U937 cells and BAEC. Total RNAs were isolated from macrophage-like differentiated U937 cells and BAEC at the indicated times after LPS (10 μg/ml) stimulation. RNAs (24 μg) fractionated through formaldehyde-containing agarose gel electrophoretically were transferred to a nylon membrane and hybridized with a ³²P-labeled COX-2, TXS (thromboxane synthase), PGIS (prostacyclin synthase), C/EBPδ, and GAPDH (glyceraldehydes-3-phosphate dehydrogenase probes), respectively.

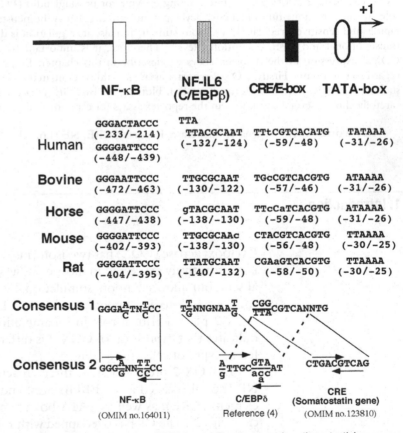

Fig. 2. Proximal promoter region of COX-2 gene. The diagrams show the potential response elements based on sequence similarities to consensus response elements. Distances are given as nucleotide positions relative to the transcriptional start site as +1.

for C/EBPδ (4), which will partly make possible the various expression patterns of COX-2. We constructed several COX-2 reporter vectors to study the roles of these *cis*-acting elements.

Fortunately, by using these reporter vectors, we and our collaborators have been showing that these *cis*-acting elements are differently involved in the COX-2 promoter activity in different cells ((5–18) (see http://koto.nara-wu.ac.jp/kenkyu/Profiles/5/0000469/theses_e1.html for more details) (see Notes 1 and 2). On the other hand, other *cis*-acting elements such as NF-AT sites (19) and PPRE (20) were reported in the COX-2 promoter region. Therefore, you may need to make your own reporters of COX-2 gene for your experimental purposes.

Transfection assays recently became more widely used. Various transfection reagents, reporter vectors, and luminometers are available from several suppliers. However, it is still difficult to perform transfection assays on some cell lines. Moreover, in the case of U937 cells, there exist two distinct cell types: one may be suitable for transfection assay but does not express COX-2, and the other type is difficult for the transfection assay but shows COX-2 induction by LPS after differentiation. In this case, we found that the nuclear receptor peroxisome proliferator activated receptor-γ (PPARγ) is only expressed in the latter type (10). Moreover, better transfection efficiency sometimes brings about more cell damage; for example, some types of cellular stress cause induction of COX-2. Therefore, one should measure COX-2 promoter activity without stimulation to determine whether COX-2 induction is due to stress or stimulation. In addition, measuring an induced promoter activity of COX-2 gene is also not easy because transfected cells need considerable time to recover from the damage of the transfection. Therefore, one should determine the appropriate conditions for the cell type in question by performing pilot experiments.

The method of reporter assays described below is for bovine arterial endothelial cells (BAEC). This method will be modified for various cell types by using different transfection reagents (see Note 3). It is also important to evaluate your luciferase assays, at least, by RT-PCR or northern blot analysis.

2. Materials

2.1. Cell Culture and Cell Lysis

1. Dulbecco's Modified Eagle's Medium (DMEM) supplemented with or without 10% fetal bovine serum (FBS, see Note 4). Store at 4°C.

2. Phosphate buffered saline (PBS) for cell culture: 137 mM of NaCl, 2.7 mM of KCl, 4.2 mM of Na_2HPO_4, 1.4 mM of KH_2PO_4. The final pH should be adjusted at 7.4. Store at 4°C.

3. Trypsin (0.25%) and EDTA (0.02%) solution (cell culture tested, sterile-filtered). Store at –20°C. The working solution (0.05% trypsin, 0.01% EDTA) is freshly prepared and stored at 4°C within several days.

4. LPS (from *Escherichia coli* serotype 055: B5) is dissolved at 1 mg/ml in tissue culture grade water, sterile-filtered and stored at 4°C. Aliquot and store at –80°C for long-term storage.

5. Reporter lysis buffer (Promega Cat. # 3971) is used for cell lysis. However, the manufacturing company does not reveal the composition of the buffer. So, if you do not want to use this buffer, the following buffer will also work: 25 mM of glycylglcine (pH7.8), 15 mM of $MgSO_4$, 4 mM of EGTA, 1%(v/v) of Triton X-100, 1 mM of dithiothreitol (DTT, add immediately before use). Store at room temperature.

6. Cell scrapers.

7. 24-well cell culture dishes. Lower transfection efficiency may need larger culture dishes such as 6-well or 12-well.

2.2. Transfection Reagents

1. Reporter vectors. Human COX-2 reporter vectors phPES2 (–327/+59) and (–1,432/+59) are available from our laboratory upon request (5). pSVβ-gal (Promega), an expression vector for β-galactosidase under control of SV40 promoter, is used for normalization of transfection. pEGFP-C1 (Clontech), expression vector for GFP (green fluorescent protein) is used for the confirmation of transfection efficiency before the reporter assay (optional). It is essential to obtain high quality plasmid DNAs for transfection assays. We usually use a plasmid purification kit from QIAGEN (see Note 5). Each plasmid DNA is dissolved in 0.1× TE (10 mM of Tris-HCl, pH 7.6, and 1 mM of EDTA), which is carefully prepared using endotoxin-free tissue culture grade water. Concentration of each DNA solution is kept at 1 mg/ml. Aliquot and store at –80°C (for long-term storage) or at 4°C (for short-term storage).

2. Trans IT-LT-1 (Mirus, Madison, WI). This lipid-mediated transfection reagent is suitable for BAEC, but not for other cells such as HEK293 cells.

3. Sterile Polystyrene Tubes (12 × 75 mm).

2.3. Reporter Assays

1. Luciferase assay reagent (Promega E3971). Store aliquots at –80°C. In order to avoid over three times of freezing/thawing, determine the appropriate aliquot size in your experiments (in our case, 3 ml).

2. β-galactosidase assay reagent: 80 mM of sodium phosphate butter, pH7.3, 102 mM of 2-mercaptoethanol, 9 mM of $MgCl_2$, 8 mM of CPRG (β-d-galactopyranoside, add immediately before use). Store this reagent without CPRG at room temperature. CPRG is dissolved at 400 mM in tissue culture water and stored as a stock solution aliquoting at –80°C.

3. Methods

3.1. Preparation of Cells for Transfection

1. In a 24-well tissue culture plate, seed 2×10^4 cells per well in 0.5 ml of DMEM containing 10% FBS.
2. Incubate the cells at 37°C in a CO_2 incubator until the cells are 60–70% confluent. This will usually take 18–24 h.

3.2. Transfection and Cell Lysis

1. Prepare the following solutions A and B in sterile polystyrene tubes. This is an example of transfection for 24 wells.
2. Solution A: Dilute 32.5 μl (0.8 μl × 26) of Trans-IT-LT-1 into 650 μl of DMEM without FBS, mix by gentle pipetting, and incubate at room temperature for 15 min.
3. Solution B: Dilute 7.8 μg (0.3μg × 26) of a luciferase reporter vector pGV-PES2(−327/+59) and 1.6 μg of β-galactosidase reporter vector pSVβ-gal (0.06 μg × 26) into 650 μl of DMEM without FBS (i.e. 25 μl per well). Prepare for 26 wells not for 24 wells to avoid shortage of the solutions, and mix by gentle pipetting. It is important that the medium contains no antibiotics such as penicillin and streptomycin. If you also use optional reporter vector such as pEGFP-C1, each DNA amount should be 0.8 μg each for pSVβ-gal and pEGFP-C1 to keep total DNA amounts constant (see Note 6).
4. Add solution B to solution A, mix gently, and incubate at room temperature for 15 min.
5. Add the mixed solutions A and B (50 μl per well), dropwise to the cells in complete growth medium in Subheading 3.1. Gently rock the dish back and forth and from side to side to distribute the DNA/lipid complexes evenly.
6. Incubate for 5 h, replace 0.5 ml of the fresh DMEM containing 10% FBS and incubate for an additional 24 h (see Note 7).
7. Add LPS (final working concentration 10 μg/ml), and incubate for an additional 5 h (see Note 8). Other Stimuli can be used along with or without LPS.
8. Observe the transfected cells with pEGFP-C1 expression vector by using a fluorescence microscope. If expression of GFP is detected in over 30% of transfected cells, it will be good for transfection assays for BAEC.
9. Remove the growth medium from the wells. Wash the cells once with PBS, being careful not to dislodge any of the cells. Remove as much of the final PBS wash as possible.
10. Add 100 μl per well of reporter lysis buffer to cover the cells. Rock the dish slowly several times to ensure complete coverage of the cells.

11. Scrape all areas of the dish, then tilt the dish and scrape the cell lysate to the lower edge of plate. Take care to scrape down all visible cell debris. Transfer the cell lysate to a microcentrifuge tube with a pipette and place the tube at −80°C for at least 15 min (see Note 9).

12. After thawing the cell lysate at room temperature, vortex the tube for 15 s, then centrifuge in a microcentrifuge ($11,000 \times g$) for 15 s at room temperature. Transfer the supernatant to a fresh tube.

13. Mix 10 µl of cell extract with 50 µl of Luciferase Assay Reagent in sample tubes at room temperature (see Note 10).

14. Place the reaction mixture in a luminometer. Measure the light produced for 10 s. Assay each reaction at the same time interval after addition of the sample to assay reagent (see Note 11).

15. Mix 2 µl of cell extract (room temperature) with 100 µl of β-galactosidase assay reagent on a microtiter plate. As a control, mix 2 µl of Reporter Lysis Buffer (room temperature) with 100 µl of β-galactosidase assay reagent. Incubate samples at room temperature for 5–30 min.

16. Read the absorbance at 580 nm with a microtiter plate reader (see Note 12).

17. Calculate the normalized luciferase activity by data obtained from steps 14 and 16. For example, when the measured values in step 14 are LucA1, LucA2, and LucA3 (without stimulus X) and LucB1, LucB2, and LucB3 (with stimulus X), respectively, and the corresponding measured values in step 16 are bGalA1, bGalA2, and bGalA3 (without stimulus X) and bGAlB1, bGalB2, and bGalB3 (with stimulus X), respectively, firstly, we obtain the values LucA1/bGalA1, LucA2/bGalA2, and LucA3/bGalA3, and LucB1/bGalB1, LucB2/bGalB2, and LucB3/bGalB3, then calculate the means and standard deviations for A and B, respectively. If these means and standard deviations are expressed by M(A), SD(A), M(B), and SD(B), respectively, the following values are expressed as

Normalized luciferase activity A (without stimulus X)
$= 1 \pm SD(A)/M(A)$

Normalized luciferase activity B (with stimulus X)
$= M(B)/M(A) \pm SD(B)/M(A)$

4. Notes

1. Human COX-1 and COX-2 reporter vectors are available for your request (7, 9: e-mal to H.I.). Luciferase activity of the COX-2 reporter is higher than that of the COX-1 reporter,

whereas COX-2 mRNA expression is much lower than COX-1 mRNA without stimulation in several cells. This discrepancy comes from COX-2 instability sequences in the 3′-untranslated region, whose reporter vector is also available (13).

2. We constructed the COX-2 reporter vectors using pGL-2 (Promega). We also constructed COX-2 reporter vectors using pGL-3 (Promega), which encodes a stable luciferase gene. However, induction of the latter type is lower than that of the former type. We also constructed COX-2 reporter vectors using several GFPs as a reporter; however, we could not measure the good dynamic range of FACS analysis using the GFP compared with the luciferase activity in the induction experiment. Experimenters should also pay attention to other assays such as Chromatin immunoprecipitation and gel-shift assays to evaluate transfection assays.

3. There are several transfection methods. In our case, we usually use lipid-mediated transfection reagents because of convenience and stable results in spite of their higher cost. In transfection into BAEC, firstly, we used a lipofectin (Invitrogen) (7), and then changed to Trans-IT-LT1 (Mirius) (9). On the other hand, in transfection into HepG2 and HEK293 cells, we use a lipofectamin (Invitrogen). Recently, many transfection reagents are available. You could carefully choose the reagents for your experiments. At that time, you had better check an induction of luciferase activity by stimulants, not basal luciferase activity without stimulants because COX-2 promoter may be activated by the reagents themselves due to their potential damaging effects to transfected cells (see Note 4).

4. Choice of a good FBS is very important for COX-2 transfection assays. We usually choose a suitable serum for COX-2 reporter assay. As an example, Fig. 3 shows different induction of the COX-2 promoter activity by LPS in BAEC cultured with four different lots of FBS. In this case, we chose Lot #2 because of lower luciferase activity without LPS and higher activity with LPS. Fig. 3a, b show luciferase activities derived from phPES2 (−327/+59) with and without normalization of β-galactosidase activities derived from pSV-β gal, respectively.

5. A special plasmid DNA purification kit which removes endotoxin is available from QIAGEN. Although this may be more effective, we have observed no difference in the results of reporter assays using plasmid DNA prepared by normal and special kits.

6. If you cotransfect an expression vector, such as PPARα, with COX-2 reporter vector, the total amount of plasmids remains constant. In this case, for example, 3.9 μg PPARα expression vector, 3.9 μg COX-2 reporter, 0.9 μg pSV-β-gal, and 0.9 μg pEGFP-C1 are used (10). If you investigate the effects of an expression vector for X protein on COX-2 promoter, you

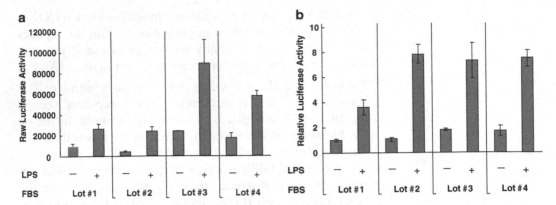

Fig. 3. Effects of different lots of FBS on COX-2 promoter activity. BAEC were transiently transfected with phPES2(−327/+59) together with pSV-βgal used as an internal control for the transfections. Following transfection, the cells were incubated for 5 h with no stimulant or with LPS (1 μg/ml) in the presence of four different lots of FBS. The cells were then harvested, lysed, and assayed for both luciferase and β-galactosidase activities as described. Results are represented as raw luciferase activities (a) or relative luciferase activities (b) obtained by dividing the normalized luciferase activity in lot #1.

should be using the same expression vector without gene for X protein as a negative control.

7. Incubation time before adding stimulants, such as LPS, should be determined by each preliminary experiment. In our case, induction of COX-2 promoter activity is not observed in the incubation for 5 h because transfected cells are exposed to the stimuli of the transfection itself and will not be ready to respond to stimulants. In the case of MC3T3E1 cells (21), an incubation of 72 h before adding stimulants is needed for good induction of the luciferase. According to our experience, longer incubation time is better for measuring the induced luciferase activities, but not for measuring the basal luciferase activities.

8. Incubation time after adding stimulants such as LPS should also be determined by performing preliminary experiments. We recommend a time-course experiment for each stimulant. Stimuli should be dissolved in DMSO or ethanol. Stock solutions should be concentrated 1,000 times more than the final concentration used in the culture medium in order to dilute DMSO and ethanol significantly.

9. Luciferase in the cell extract is not stable even if you store it at −80°C. On the other hand, a freezing/thawing cycle provides better extraction of cells. We recommend measuring the luciferase activity using freshly prepared samples after a freezing/thawing cycle. If you use the stored sample, you should measure all samples in the set because a luciferase activity measured on one day is different from that measured on another day.

10. Measuring the luciferase activity is sensitive to temperature. It is important to keep the samples at room temperature. The luminometer should be turned on at least 15 min before use.

11. Luciferase activity of each transfected sample should be at least three times the negative control without nontransfected cell extract. Dynamic range and linearity of luciferase activity should be determined by each luminometer.

12. The absorbance of each transfected sample should be at least three times the negative control without nontransfected cells. Dynamic range and linearity of β-galactosidase activity should be determined by each plate reader. In general, dynamic range of β-galactosidase activity is much narrower than that of luciferase activity.

Acknowledgments

We would like to thank Ms. Haruka Takeuchi, Mariko Hotta, Tomoko Tsukamoto, and Tomomi Matsuyama for their technical assistance. This work was supported by Grant-in-Aid for scientific research from Ministry of Education, Culture, Sports, Science and Technology (Houga18650213, B19300250).

References

1. Kang YJ, Mbonye UR, DeLong CJ, Wada M, Smith WL (2007) Regulation of intracellular cyclooxygenase levels by gene transcription and protein degradation. Prog Lipid Res 46:108–125

2. Dannenberg AJ, Subbaramaiah K (2003) Targeting cyclooxygenase-2 in human neoplasia: rationale and promise. Cancer Cell 4:431–436

3. Sirois J, Sayasith K, Brown KA, Stock AE, Bouchard N, Dore M (2004) Cyclooxygenase-2 and its role in ovulation: a 2004 account. Hum Reprod Update 10:73–85

4. Osada S, Yamamoto H, Nishihara T, Imagawa M (1994) DNA binding specificity of the CCAAT/enhancer-binding protein transcription factor family. J Biol Chem 271:3891–3896

5. Inoue H, Nanayama T, Hara S, Yokoyama C, Tanabe T (1994) The cyclic AMP response element plays an essential role in the expression of the human prostaglandin-endoperoxide synthase 2 gene in differentiated U937 monocytic cells. FEBS Lett 350:51–54

6. Nanayama T, Hara S, Inoue H, Yokoyama C, Tanabe T (1995) Regulation of two isozymes of prostaglandin endoperoxide synthase and thromboxane synthase in human monoblastoid cell line U937. Prostaglandins 49:371–382

7. Inoue H, Yokoyama C, Hara S, Tone Y, Tanabe T (1995) Transcriptional regulation of human prostaglandin-endoperoxide synthase-2 gene by LPS and phorbol ester in vascular endothelial cells. J Biol Chem 270:24965–24971

8. Inoue H, Tanabe T (1998) Transcriptional role of the nuclear factor κB site in the induction by LPS and suppression by dexamethasone of cyclooxygenase-2 in U937 cells. Biochem Biophys Res Commun 244: 143–148

9. Inoue H, Umesono K, Nishimori T, Hirata Y, Tanabe T (1999) Glucocorticoid-mediated suppression of the promoter activity of cyclooxygenase-2 gene is modulated by expression of its receptor in vascular endothelial cells. Biochem Biophys Res Commun 254:292–298

10. Inoue H, Tanabe T, Umesono K (2000) Feedback control of COX-2 expression through PPARγ. J Biol Chem 275:28028–28032

11. Inoue H, Taba Y, Miwa Y, Yokota C, Miyagi M, Sasaguri T (2002) Transcriptional and post-transcriptional regulation of cyclooxygenase-2 expression by fluid shear stress in vascular endothelial cells. Artherioscler Thromb Vasc Biol 22:1415–1420

12. Nie N, Pang L, Inoue H, Knox AJ (2003) Transcriptional regulation of cyclooxygenase 2 by bradykinin and interleukin-1beta in human airway smooth muscle cells. Mol Cell Biol 24:9233–9244

13. Nishimori T, Inoue H, Hirata Y (2004) Involvement of the 3′-untranslated region of cyclooxygenase-2 gene in its post-transcriptional regulation throught the glucocorticoid receptor. Life Sci 74:2505–2513

14. Norata GD, Callegari E, Inoue H, Catapano AL (2004) HDL3 induces cyclooxygenase-2 expression and prostacyclin release in human endothelial cells via a p38 MAPK/CRE-dependent pathway. Artherioscler Thromb Vasc Biol 24:871–877

15. Chen J, Zhao N, Rao R, Inoue H, Hao CM (2005) C/EBPβ and its binding element are required for NFκB induced COX2 expression following hypertonic stress. J Biol Chem 280:16354–16359

16. Park SW, Sung MW, Heo DS, Inoue H, Shim SH, Kim KH (2005) Nitric oxide upregulates the cyclooxygenase-2 expression through the cAMP-response element in its promoter in several cancer cell lines. Oncogene 24:6689–6698

17. Kuwamoto S, Inoue H, Tone Y, Izumi Y, Tanabe T (1997) Inverse gene expression of prostacyclin and thromboxane synthases in resident and activated peritoneal macrophages. FEBS Lett 409:242–246

18. Ricote M, Li AC, Willson TM, Kelly CJ, Glass CK (1998) The peroxisome proliferator-activated receptor-gamma is a negative regulator of macrophage activation. Nature 391:79–82

19. Pontsler AV, Hilaire A, Marathe GK, Zimmerman GA, McIntyre TM (2002) Cyclooxygenase-2 is induced in monocytes by peroxisome proliferator activated receptor gamma and oxidized alkyl phospholipids from oxidized low density lipoprotein. J Biol Chem 277:13029–13036

20. Iñiguez MA, Martinez-Martinez S, Punzón C, Redondo JM, Fresno M (2000) An essential role of the nuclear factor of activated T cells in the regulation of the expression of the cyclooxygenase-2 gene in human T lymphocytes. J Biol Chem 275:23627–23635

21. Yamamoto K, Arakawa T, Ueda N, Yamamoto S (1995) Transcriptional roles of nuclear factor kappa B and nuclear factor-interleukin-6 in the tumor necrosis factor alpha-dependent induction of cyclooxygenase-2 in MC3T3-E1 cells. J Biol Chem 270:31315–31320

Chapter 3

Purification of Recombinant Human COX-1 and COX-2

James K. Gierse

Abstract

Human recombinant COX-1 and COX-2 was prepared in a purified form. The genes were cloned and expressed in insect cells, extracted by detergents, and purified by ion-exchange followed by size exclusion chromatography. Insect cells from 10 L fermentation yielded 2.3 mg of COX-1 with an overall yield of 75%, and 29 mg of COX-2 with an overall yield of 45%. Enzyme prepared in this manner was fully active and proved to be useful in biophysical studies including direct binding studies. The COX-2 provided material that was subsequently used in X-ray crystallography studies.

Key words: COX-2, Cyclooxygenase, PGH2 synthase, Prostaglandin, Prostaglandin endoperoxide synthase

1. Introduction

Cyclooxygenase is a bi-functional enzyme that first catalyzes the addition of two moles of molecular oxygen to arachidonic acid to form the hydroperoxide PGG2, then reduces the hydroperoxide to the alcohol, PGH2, by a peroxidase activity (1). PGH2 is short-lived in aqueous solutions and is quickly reduced to a mixture of prostaglandins, depending on the downstream synthase found, or in the absence of synthases, to a mixture of PGE2/PGD2 (2). Cyclooxygenase is found in two forms: COX-1, which is constitutively expressed in most cells, is responsible for the production of prostaglandins that maintain homeostasis; and COX-2, which is upregulated in inflammatory cells in response to an inflammatory stimulus (cytokines, LPS, etc.), is responsible for the production of prostaglandins at the site of inflammation (3–6). COX-2 is also expressed to a lesser extent constitutively in brain and kidney (7). Non-steroidal anti-inflammatory drugs (NSAIDs) act by inhibiting the production of prostaglandins at the site of

Samir S. Ayoub et al. (eds.), *Cyclooxygenases: Methods and Protocols*, Methods in Molecular Biology, vol. 644, DOI 10.1007/978-1-59745-364-6_3, © Springer Science+Business Media, LLC 2010

inflammation by blocking the oxygenase site of cyclooxygenase. The advantage of a COX-2 selective inhibitor soon became apparent, and testing schemes were developed that relied on using partially purified enzymes expressed in insect cells (8). These in vitro assays provided a powerful means for assessing COX selectivity and potency and led to the discovery and clinical development of the first rationally designed COX-2 selective inhibitors, celecoxib (9, 10) and rofecoxib (11).

Purified enzymes have been useful for defining the kinetic mechanism of COX-2 selective inhibition and for structural studies of the molecular basis underlying this phenomenon. The X-ray crystal structure of COX-1 was solved utilizing enzyme purified from sheep seminal vesicle (12). X-ray crystal structures of COX-2 were determined using the murine form of recombinant COX-2 (13). The purification scheme outlined here was optimized for production of high-quality crystals that could be used for structural work and was utilized to produce both human and murine COX-2. Purified proteins allowed for the development of biophysical methods that were useful in the elucidation of the mechanism of inhibition. Among these were kinetic studies utilizing oxygen consumption (14) and direct binding studies that were useful in measuring on and off rates of compounds (15).

2. Materials

2.1. Detergents

(1) n- Decyl-L-D-maltoside (C10M) (2) ß-octyl-glycoside (BOG) (Calbiochem, La Jolla, CA).

2.2. Buffers

Note: Quantities are given for 10 L fermentation prep, filter all buffers with 0.45 µ filter, and store at 4°C. All buffer components were purchased from Sigma Chemical (St. Louis, MO).

2.2.1. Buffer Stocks

1. 1 M Tris-HCl pH 8.1 (at room temp): 121 g Tris base + 0.7 L dH$_2$O titrate to pH 8.1 w/concentrated HCl bring to final volume of 1 L with H$_2$O.
2. 0.2 M Sodium EDTA: 83.2 g Tetrasodium EDTA/1 L dH$_2$O, titrate pH to 8 with concentrated liquid NaOH.
3. 0.5 M Diethyldithiocarbamate (DEDTC): 0.0855 g/10 ml, store frozen in 1 ml aliquots (add fresh to buffers immediately before use).
4. 5 M NaCl: 292 g/1,000 ml dH$_2$O.

2.2.2. Buffers

1. *Homogenization buffer:* 25 mM Tris-HCl pH 8.1, 0.25 M Sucrose, 1 mM EDTA, 1 mM DEDTC. To make 500 ml mix 12.5 ml 1 M Tris-HCl pH 8.1, 42.8 g sucrose, 2.5 ml of 0.2 M EDTA, and 1 ml of 0.5 M DEDTC.

2. *Preequilibrium buffer for Q column*: 500 mM Tris-HCl pH 8.1, 0.5 M NaCl. To make 250 ml mix 125 ml 1 M Tris-HCl pH 8.1 and 25 ml 5 M NaCl.

3. *Buffer A (Equilibrium buffer) for Q column*: 20 mM Tris-HCl pH 8.1, 0.1 mM EDTA, 0.1 mM DEDTC, 0.15% (w/v) Decyl-L-D-maltoside. To make 1,000 ml mix 20 ml 1 M Tris-HCl pH 8.1, 0.5 ml of 0.2 M EDTA, 0.2 ml of 0.5 M DEDTC (added just prior to use), and 1.5 g solid C10M.

4. *Buffer B (Elution buffer) for Q column*: 20 mM Tris-HCl pH 8.1, 0.3 M NaCl, 0.1 mM EDTA, 0.1 mM DEDTC, 0.15% Decyl-L-D-maltoside. To make 500 ml mix 10 ml 1 M Tris-HCl pH 8.1, 30 ml of 5 M NaCl, 0.25 ml of 0.2 M EDTA and 0.1 ml of 0.5 M DEDTC (added just prior to use) and 1.5 g solid C10M.

5. *Size exclusion buffer*: 25 mM Tris-HCl, 0.1 mM EDTA, 0.1 mM DEDTC, 0.3% ß-octyl-glycoside. To make 1,000 ml mix 25 ml 1 M Tris-HCl pH 8.1, 0.5 ml of 0.2 M EDTA and 0.2 ml of 0.5 M DEDTC (added just prior to use), and 3 g solid BOG.

2.3. Columns

Note: All chromatography solutions should be degassed prior to detergent addition.

1. Anion-exchange Column #1 (AIEX #1) 2.6×10 cm (50 ml) with a packing material of Macroporous Q (Bio Rad, Hercules, CA), wash with 50–100 ml 500 mM Tris/0.5 M NaCl pH 8.1, followed by buffer A without detergent, then 50–100 ml buffer A with detergent at a flow rate of 10 ml/min (120 linear cm/h).

2. Size Exclusion Column (SEC): 2.6×90 cm (500 ml) with a packing material of Sephacryl S-200 HR (Amersham-Pharmacia, Piscataway, NJ). Equilibrate column with 750 ml size exclusion buffer at a flow rate of 3 ml/min (30 linear cm/h).

3. Anion Exchange Column #2 (AIEX #2): Column: 8 ml pre-packed, Mono-Q (Amersham-Pharmacia, Piscataway, NJ). Equilibrate with 2–3 column volumes 10× buffer A followed by 10 column volumes buffer B.

2.4. Concentration Membranes

1. Amicon YM-30 76 mM (Millipore, Billerica, MA).

2. Soak membrane in water for 30 min.

3. Put concentrator together and pressure test with water at 50 PSI.

2.5. SDS-Polyacrylamide Gel Electrophoresis (SDS-PAGE)

1. Gels: 10–20% Tricine Gel 1.0 mm×12 well (Invitrogen, Carlsbad, CA).

2. Sample buffer: 2× Tris-Glycine SDS (Invitrogen, Carlsbad, CA).

3. BME: 2-mercaptoethanol (Sigma Chemical, St. Louis, MO).

4. Running buffer: Tris-Glycine SDS (Invitrogen, Carlsbad, CA).

5. Stain: Colloidal blue (Invitrogen, Carlsbad, CA).

2.6. Activity Assay

1. TMPD: N,N,N',N'-tetramethyl-p-phenylenediamine (Sigma Chemical, St. Louis, MO). 17 mM Stock 4 mg/ml in dH_2O. Store 1 ml aliquots frozen, keep on ice. Use once, discard unused. (TMPD quickly oxidizes at room temp.)

2. Heme: Bovine hemin Sigma Chemical, St. Louis, MO) 100 µm stock in DMSO. Store frozen in 0.5 ml aliquots.

3. Arachidonic acid sodium salt (NuChek Prep, Elysian, MN) 10 mM stock in dH_2O, store frozen in 1 ml aliquots.

4. Assay conditions: 100 mM Tris-HCl pH 8.1, 1 µM Heme, 100 µM Arachidonic acid 170 µM TMPD, room temperature, read plate at 590 nm after 10 min.

5. Assay buffer: To prepare assay buffer mix 16.96 ml H_2O, 2.2 ml 1 M Tris-HCl pH 8.1, 2.2 ml 100 µM Heme, 0.22 ml 10 mM Na arachidonate and 0.22 ml 17 mM TMPD.

2.7. Protein Assay

1. Coomassie Plus Protein Assay (Pierce, Rockford, IL).

2. BSA standard (2 mg/ml).

3. Methods

All operations carried out at 4°C unless otherwise noted.

3.1. Wash Cells (Day 1)

1. Suspend frozen cells in 10 volumes homogenization buffer without detergent. (20 g wet weight cells + 200 ml Buffer).

2. Stir until suspended at room temp 1/2 h.

3. Centrifuge 20 min at 10,000×g (Beckman JA-13 rotor @ 8,000 rpm).

4. Discard Supernatant (add bleach to 1% to disinfect).

3.2. Solubilisation

1. Resuspend cells in 10 volumes (w/v) homogenization buffer.

2. Add decyl-maltoside (C10M) to 1.5% final drop wise from a 15% stock solution (225 ml + 25 ml 15% stock). Stir for 1 h.

3. Centrifuge at 25,000×g (Beckman JA-13 rotor 13,000 rpm) for 20 min, discard Pellet with addition of 1% bleach.

4. Assay supernatant for protein and peroxidase activity (2–20 µl in microtiter assay).

Save sample for SDS-PAGE (50 µl sample + 50 µl 2× sample buffer + 5 µl BME; Boil 5 min) (Note 1)

3.3. Anion-Exchange Column (50 ml)

1. Load detergent extract directly onto equilibrated Q column, collect flow through for assay.

2. Wash with 200 ml equilibration buffer (or until UV baseline returns to 0), collect wash.

3. 1 L linear gradient of 500 ml A + 500 ml B, collect 10 ml fractions.

4. Assay load, flow through, wash and 5 μl samples from every other fraction in microtiter assay.

Pool fractions with 90% of activity, discard head and tail fractions, (Pool vol. 200–300 ml) and assay pool for protein (Bradford) and peroxidase activity (2–20 μl in microtiter assay), save sample for SDS-PAGE (Fig. 1).

3.4. Membrane Concentration (Day 2)

1. Concentrate Q-column pool to decrease volume for sizing column with YM-30 membrane with a PSI of 20 to final volume of 10 ml.

2. Wash membrane with 5 ml sizing column buffer.

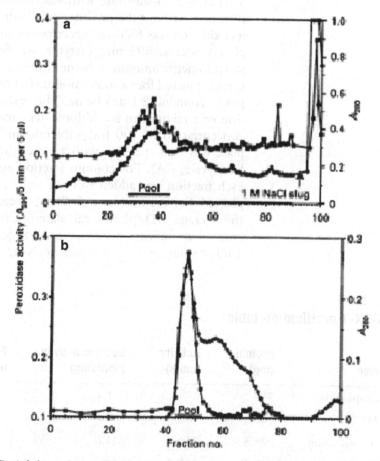

Fig. 1. Anion exchange and size exclusion chromatography for COX-2 Purification. X axis is fraction number and Y axis is OD 590 of the TMPD activity assay and OD @ 280 nm.

3.5. Size-Exclusion Column (500 ml)

1. Load 15–20 ml Q column concentrate on to 500 ml SEC column (3–4% of column volume = load volume).
2. Column running conditions: Flow Rate, 0.88 ml/min, AUFS: 2.0, Fraction size: 5 ml (100 fractions).
3. Assay 5 µl samples of each fraction using microtiter assay, pool main peak.
4. Assay pool for protein and activity, save sample for SDS-PAGE.

3.6. Anion-Exchange #2 (necessary for COX-1) Day 3

1. Dilute SEC pool with 10× buffer B to reduce iconic strength.
2. Load onto 8 ml Mono Q column 2 1 ml/min.
3. Run linear gradient of 20 column volumes (160 ml) buffer B to C.
4. Collect 5 ml Fractions.
5. Assay 5 µl samples.

3.7. Activity Assay

The cyclooxygenase activity was indirectly measured by utilizing TMPD as a co-substrate with arachidonic acid. TMPD will not turn over without the presence of a hydroperoxide substrate and its oxidation was followed spectrophotometrically with a 96 well plate reader at 595 nm. Enzyme was first reconstituted with a stoichiometric amount of heme. Enzyme (1 µg/well) and inhibitor were mixed for various amounts of time in 100 mM Tris-HCl, pH 8, containing 1 µM heme. The reaction was started by addition of a mixture of arachidonic acid and TMPD to give a final concentration of 100 µM substrate and 170 µM TMPD. The plate was read at 5 min with a plate reader (Molecular Devices, Sunnyvale, CA). For column fraction assays either 1 or 5 µl of each fraction was added to each well. The raw OD @595 were plotted. For the final assay 2–20 µl of each pool was assayed and the average OD/µl was calculated. Units were defined as the amount of enzyme necessary to change the OD 595 1 unit in 1 ml of solution (Tables 1 and 2) (Note 2).

Table 1
COX-1 purification table

Step	Protein (mg)	Activity (units)	Specific activity (units/mg)	Fold purification	Yield (%)
Chaps extract	792	858	1.1		
AIEX	133	982	7.4	6.7	114
Size exclusion	35.5	768	21.6	20	90
Second AIEX	2.3	645	280	245	75

3.8. Final Protein Assay

1. Coomassie Plus (Pierce #23236): Microtiter method (96 Well).

2. Standard Curve: BSA, 2 mg/ml. Dilute to 0.02 µg/µl, add 0.2–1.6 µg in 100 µl

3. Samples: 100 µl sample + H_2O, 3× serial dilution.

4. 100 µl Coomassie Plus Reagent.

5. Read OD @ 590.

6. Calculate protein concentration of samples that fall in the middle of the standard curve.

Table 2
COX-2 purification table

Step	Protein (mg)	Activity (units)	Activity (units/mg)	Fold purification	Yield (%)
Chaps extract	1,001	13,958	14		
AIEX	58	8,990	155	11	64
Size exclusion	29	6,840	226	16	46

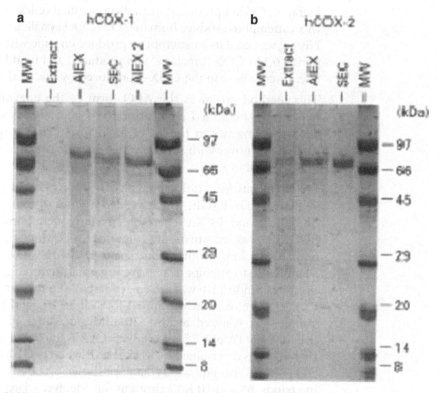

Fig. 2. SDS-PAGE of purification steps for recombinant human COX-1 and COX-2. 2 µg total protein were loaded per lane on a 10–20% 10 × 10 cm mini-gel and stained with colloidal Coomassie.

3.9. Storage

1. Aliquot into polypropylene tubes and quick freeze in dry ice/ methanol for long-term storage.

2. Enzyme to be used for crystallography was stored at 4°C for up to 1 week.

3.10. SDS-PAGE (Note 3)

1. 2 μg total protein per lane, Sample from each step, 10–20% 10 × 10 cm mini-gel.
   Colloidal Coomassie stain (Fig. 2).

---

4. Notes

1. This protocol was optimized for the production of COX-2 for crystallography. Optimization was carried out for the detergents used, their concentration, and the timing of the process. Detergents that can be substituted for decyl-maltoside for solubilization are chaps, tween-20 and triton X-100, with the last two detergents not very amenable to crystallography. The COX-1 produced by this method was not high-quality crystallography. The yields were approximately 10 times lower and the overall purity somewhat lower than COX-2. For X-ray crystallography, COX-1 produced from sheep seminal vesicle was used. In an attempt to produce human COX-1 for crystallography, a 6 His tag was used in an attempt to produce enzyme with greater purity. 6-His COX-2 material was produced and found to have inferior crystals than the COX-2 produced by this method.

2. This method produces the APO form of the enzymes, and heme must be added back prior to assay. It is advisable to keep the enzymes in the APO form for storage to prevent catalytic turnover with substrate that may be present in membrane preparations.

3. Heterogeneity from differential glycosylation of the sample was shown by SDS-PAGE, which yielded a series of four bands at 68, 70, 72, and .74 Kd. Therefore, in subsequent purifications, hCOX-2 was enzymatically digested following the anion exchange column with recombinant Endo-H-F (Boehringer Mannheim) to attempt to remove sugars at the three glycosylation sites. Endo-H-F was added at a rate of 0.6 mg per 1 mg of COX-2. The reaction continued for 6 h at 4°C and then the sample was dialyzed against 20 mM potassium phosphate, 0.1 mM EDTA pH 7.5 and 0.3% β-OG. Removal of sugars at two sites was determined by SDS-PAGE when two bands at 72 and 74 Kd were eliminated (data not shown). The two remaining bands, 68 and 70 Kd, represent fully deglycosylated protein and protein with one intact glycosylation site, respectively.

Acknowledgments

The author would like to thank Scott Hauser and Dave Creely for cloning human COX-1 and COX-2 and Shaukat Rangwala for construction of the baculovirus vectors and Jennifer Pierce for fermentation.

References

1. Smith WL, Song I (2002) The enzymology of prostaglandin endoperoxide H synthases-1 and -2. Prostaglandins Other Lipid Mediat 68–69:115–128

2. Jakobsson PJ, Thoren S, Morgenstern R, Samuelsson B (1999) Identification of human prostaglandin E synthase: a microsomal, glutathione-dependent, inducible enzyme, constituting a potential novel drug target. Proc Natl Acad Sci U S A 96(13):7220–7225

3. Mitchell JA, Akarasereenont P, Thiemermann C, Flower RJ, Vane JR (1993) Selectivity of nonsteroidal antiinflammatory drugs as inhibitors of constitutive and inducible cyclooxygenase. Proc Natl Acad Sci U S A 90(24): 11693–11697

4. Masferrer JL, Seibert K (1994) Regulation of prostaglandin synthesis by glucocorticoids. Receptor 4(1):25–30

5. Seibert K, Masferrer JL (1994) Role of inducible cyclooxygenase (COX-2) in inflammation. Receptor 4(1):17–23

6. Crofford LJ (1997) COX-1 and COX-2 tissue expression: implications and predictions. J Rheumatol Suppl 49:15–19

7. Seibert K, Zhang Y, Leahy K, Hauser S, Masferrer J, Isakson P (1997) Distribution of COX-1 and COX-2 in normal and inflamed tissues. Adv Exp Med Biol 400A:167–170

8. Gierse JK, Hauser SD, Creely DP, Koboldt C, Rangwala SH, Isakson PC, Seibert K (1995) Expression and selective inhibition of the constitutive and inducible forms of human cyclooxygenase. Biochem J 305(Pt 2):479–484

9. Penning TD, Talley JJ, Bertenshaw SR, Carter JS, Collins PW, Docter S, Graneto MJ, Lee LF, Malecha JW, Miyashiro JM, Rogers RS, Rogier DJ, Yu SS, Anderson GD, Burton EG, Cogburn JN, Gregory SA, Koboldt CM, Perkins WE, Seibert K, Veenhuizen AW, Zhang YY, Isakson PC (1997) Synthesis and biological evaluation of the 1, 5-diarylpyrazole class of cyclooxygenase-2 inhibitors: identification of 4-[5-(4-methylphenyl)-3-(trifluoromethyl)-1H-pyrazol-1-yl]benzenesulfonamide (SC-58635, celecoxib). J Med Chem 40(9):1347–1365

10. Gierse JK, Koboldt CM, Walker MC, Seibert K, Isakson PC (1999) Kinetic basis for selective inhibition of cyclo-oxygenases. Biochem J 339(Pt 3):607–614

11. Black WC, Brideau C, Chan CC, Charleson S, Chauret N, Claveau D, Ethier D, Gordon R, Greig G, Guay J, Hughes G, Jolicoeur P, Leblanc Y, Nicoll-Griffith D, Ouimet N, Riendeau D, Visco D, Wang Z, Xu L, Prasit P (1999) 2, 3-Diarylcyclopentenones as orally active, highly selective cyclooxygenase-2 inhibitors. J Med Chem 42(7): 1274–1281

12. Picot D, Loll PJ, Garavito RM (1994) The X-ray crystal structure of the membrane protein prostaglandin H2 synthase-1. Nature 367(6460):243–249

13. Kurumbail RG, Stevens AM, Gierse JK, McDonald JJ, Stegeman RA, Pak JY, Gildehaus D, Miyashiro JM, Penning TD, Seibert K, Isakson PC, Stallings WC (1996) Structural basis for selective inhibition of cyclooxygenase-2 by anti-inflammatory agents. Nature 384(6610):644–648

14. Walker MC, Kurumbail RG, Kiefer JR, Moreland KT, Koboldt CM, Isakson PC, Seibert K, Gierse JK (2001) A three-step kinetic mechanism for selective inhibition of cyclo-oxygenase-2 by diarylheterocyclic inhibitors. Biochem J 357(Pt 3):709–718

15. Hood WF, Gierse JK, Isakson PC, Kiefer JR, Kurumbail RG, Seibert K, Monahan JB (2003) Characterization of celecoxib and valdecoxib binding to cyclooxygenase. Characterization of celecoxib and valdecoxib binding to cyclooxygenase. Mol Pharmacol 63(4):870–877

Chapter 4

Cloning and Expression of Cyclooxygenase-1 and Cyclooxygenase-2

Nicholas R. Staten and Beverly A. Reitz

Abstract

Cyclooxygenase (COX) enzymes play important roles in normal physiology and during inflammation of various cells and tissues. In order to help understand the functions of these enzymes, their genes can be cloned to facilitate the production of the proteins in recombinant form. We outline a method to clone the genes from a human macrophage cell line for expression in an insect cell line infected with recombinant baculovirus encoding these enzymes.

Key words: Cyclooxygenase (COX), Messenger ribonucleic acid (mRNA), Polymerase chain reaction (PCR), Baculovirus, Bacmid, Insect cells

1. Introduction

The cyclooxygenase (COX) enzymes catalyze the conversion of arachidonic acid to prostaglandin H_2. There are two known COX enzymes: the constitutively expressed COX-1, which seems to have a house-keeping role in most tissues, and the inducible COX-2, which is up-regulated in various inflamed tissues and cells. Due to the expression of each of these enzymes, they can easily be cloned from various tissues and cell types.

Since the early 1990s the COX enzymes have been expressed in heterologous systems such as mammalian and insect cells with varying degrees of success. COX-1 was cloned first and expressed (1) in insect cells followed soon by COX-2 (2). COX-1 protein and activity levels have typically been significantly lower than that of COX-2 (3, 4). The recombinant forms of these enzymes have been extensively studied and compared favorably to the activities of the native enzymes (5–7). In recent years, most laboratories

Samir S. Ayoub et al. (eds.), *Cyclooxygenases: Methods and Protocols*, Methods in Molecular Biology, vol. 644, DOI 10.1007/978-1-59745-364-6_4, © Springer Science+Business Media, LLC 2010

have relied upon baculovirus-infected insect cells as the system of choice for expressing both COX-1 and COX-2 enzymes.

2. Materials

2.1. Equipment

1. PCR Thermocycler.
2. Ultraviolet light box.
3. Cedex (Innovatis, Bielefeld, Germany).
4. 6-well Tissue Culture Dishes.
5. Disposable Erlenmeyer Flasks with Vented Caps.

2.2. Reagents

1. TRIzol® Reagent (Invitrogen, Carlsbad, CA).
2. Chloroform.
3. Isopropyl alcohol.
4. RNase-free water.
5. 70% and 95–100% ethanol (EtOH).
6. Synthetic PCR cloning primers.
7. 10× DNA Gel Loading Buffer.
8. E-Gel® 1.2% Pre-cast Agarose gels (Invitrogen, Carlsbad, CA).
9. DNA Molecular Weight (MW) markers.
10. BamHI and NotI Restriction Enzymes.
11. pFastBac1 (Invitrogen, Carlsbad, CA).
12. Shrimp Alkaline Phosphatase.
13. High efficiency chemically competent *E. coli*.
14. LB-carbenicillin agar plates.
15. Terrific Broth (TB).
16. 4 mm Glass Beads.
17. Carbenicillin.
18. DH10Bac cells (Invitrogen, Carlsbad, CA).
19. TGKIB = agar plates containing 10 µg/ml tetracycline, 7 µg/ml gentamicin, 50 µg/ml kanamycin, 40 µg/ml IPTG, and 300 µg/ml Bluo-gal (Invitrogen, Carlsbad, CA), or same antibiotics in S-Gal/LB Agar Blend with IPTG (Sigma, St. Louis, MO).
20. M13 Forward and Reverse Primers.
21. Sf-900 II SFM (Invitrogen, Carlsbad, CA).
22. Sf9 Cells (Invitrogen, Carlsbad, CA).
23. Cellfectin Reagent (Invitrogen, Carlsbad, CA).

24. Agarose Plaquing Medium: 3 parts 1.3× Sf-900 II plus 1 part 4% Agarose Gel plus 1× Antibiotic/Antimycotic (Invitrogen, Carlsbad, CA).

25. MTT dye: 5 mg/ml Thiazolyl Blue Tetrazolium Bromide in PBS; diluted to 1× (1 mg/ml) just before use (Sigma, St. Louis, MO).

2.3. Kits

1. Poly(A)Purist™ kit (Ambion, Austin, TX).

2. SuperScript™ III One-Step RT-PCR System with Platinum® Taq DNA Polymerase (Invitrogen, Carlsbad, CA).

3. QIAprep® PCR Purification kit (Qiagen, Valencia, CA).

4. QIAquick® Gel Extraction Kit (Qiagen, Valencia, CA).

5. Rapid DNA Ligation kit (Roche, Indianapolis, IN).

6. QIAprep® Miniprep Kit (Qiagen, Valencia, CA).

3. Methods

3.1. COX-1 and COX-2 Cloning

3.1.1. Total RNA Purification via TRIZOL Extraction

1. Homogenize 10^7 lipopolysaccharide (LPS)-stimulated human U937 macrophages in 1 ml TRIzol by repeated pipetting. The cells were treated with LPS (100 ng/ml) for 8 h prior to harvesting and storing at –80°C.

2. Transfer homogenate to a 1.5 ml tube and replace cap.

3. Incubate for 5 min at room temperature (RT) to complete the breakdown of the tissue.

4. Add 0.2 ml of chloroform, and re-secure cap.

5. Shake vigorously by hand for 15 s and incubate at RT for 3 min.

6. Centrifuge the sample at no more than $12,000 \times g$ for 15 min at 4°C.

7. Following centrifugation, transfer the upper aqueous phase to a fresh 1.5 ml tube.

8. Precipitate aqueous phase by adding 0.5 ml isopropyl alcohol and shaking.

9. Incubate at RT for 10 min and centrifuge at $12,000 \times g$ for 10 min at 4°C.

10. Carefully remove and discard the supernatant, avoiding the gel-like RNA pellet.

11. Wash the RNA pellet once with 1 ml of 75% EtOH (pre-chilled at 4°C).

12. Mix the sample by vortexing and centrifuge at $7,500 \times g$ for 5 min at 4°C.

13. Carefully remove the supernatant and discard.

14. Allow the pellet to air-dry or until there is no visible liquid remaining.

15. Dissolve RNA with 20 µl of RNase-free water by pipetting up and down.

16. Incubate for 10 min at 55°C.

17. Chill in ice bath.

18. Extracted total RNA sample can now be frozen at –80°C if not used immediately.

3.1.2. mRNA Purification via Poly(A)Purist™ Kit

1. Add 2 µl 5 M ammonium acetate, 1 µl glycogen, and 50 µl 95% EtOH to total RNA.

2. Mix thoroughly by vortexing, and place on dry ice for 30 min to precipitate RNA.

3. Recover RNA by centrifugation at $12,000 \times g$ for 30 min at 4°C.

4. Discard supernatant with careful aspiration using a fine-tipped pipette.

5. Centrifuge the tube briefly, and aspirate any additional fluid.

6. Add 1 ml 70% EtOH (pre-chilled to –20°C), and centrifuge RNA for 10 min at $12,000 \times g$ at 4°C.

7. Carefully remove the supernatant as described above (steps 4 and 5).

8. Allow to air-dry 10 min at RT.

9. Re-suspend RNA in 0.25 ml nuclease-free water by pipetting up and down five times.

10. Add 0.25 ml 2× Binding Solution and mix thoroughly.

11. Add RNA sample to 1 tube Oligo(dT) Cellulose, and mix well by inversion.

12. Heat the mixture for 5 min at 65°C.

13. Place tube on a platform shaker, and rock gently for 60 min at RT.

14. Centrifuge at $3,000 \times g$ for 3 min at RT, and aspirate to remove supernatant.

15. Place the RNA Storage Solution in a 70°C water bath to pre-warm.

16. Add 0.5 ml Wash Solution 1 to the cellulose pellet, and mix briefly by vortexing.

17. Place a spin column into a 2 ml tube, and transfer suspension to the spin column.

18. Centrifuge at $3,000 \times g$ for 3 min at RT.

19. Discard the filtrate from the tube, and place the spin column back into the tube.

20. Add 0.5 ml Wash Solution 1 to the column, replace the cap, and vortex briefly.

21. Centrifuge at $3,000 \times g$ for 3 min at RT.

22. Discard the filtrate from the tube, and place the spin column back into the tube.

23. Add 0.5 ml Wash Solution 2 to the column, replace the cap, and vortex briefly.

24. Centrifuge at $3,000 \times g$ for 3 min at RT.

25. Discard the filtrate from the tube, and place the spin column back into the tube.

26. Repeat Wash Solution 2 steps, immediately above, for two additional washes.

27. Transfer the Spin Column into a new centrifuge tube.

28. Add 200 µl of warm (70°C) the RNA Storage Solution to the column resin.

29. Cap the tube, and vortex briefly to thoroughly mix.

30. Centrifuge at $5,000 \times g$ for 2 min at RT to elute the mRNA from the spin column.

31. Leave the column in the tube and repeat elution steps immediately above with a second 200 µl aliquot of warm the RNA Storage Solution.

32. Discard the spin column.

33. To precipitate the RNA, add 40 µl 5 M ammonium acetate, 1 µl glycogen, and 1.1 ml 95% EtOH (–20°C), mix, and place on dry ice for 30 min (or –80°C, overnight).

34. Pellet the mRNA by centrifugation at $12,000 \times g$ for 30 min at 4°C.

35. Carefully remove and discard the supernatant.

36. Centrifuge the tube briefly a second time, and aspirate any additional fluid that collects with a fine-tipped pipette.

37. Add 1 ml 70% EtOH (–80°C), and vortex the tube to wash excess salts.

38. Pellet the mRNA by centrifugation for 10 min at $12,000 \times g$ at 4°C.

39. Remove supernatant carefully as described above.

40. Dissolve the mRNA pellet in 25 µl the RNA Storage Solution by pipetting.

41. Store the RNA at –80°C if not used immediately.

3.1.3. SuperScript Reverse Transciptase-Polymerase Chain Reaction (RT-PCR) Cloning

1. Prepare and label two 0.2 ml, nuclease-free, thin-walled PCR tubes on ice containing 25 μl 2× SuperScript™ Reaction Mix, 2 μl SuperScript™ III RT/Platinum® Taq Mix, 2 μl U937 mRNA and 19 μl of water in addition to the reagents below:

   COX-1: 1 μl each (10 μM) COX-1 primer pair (Table 1).

   COX-2: 1 μl each (10 μM) COX-2 primer pair (Table 1).

2. Gently mix and make sure that all the components are at the bottom of the tube.

3. Centrifuge briefly if needed.

4. Place the reaction in the pre-heated thermocycler programmed as described below for both reactions:

   (a) Reverse transcriptase (cDNA synthesis): 1 cycle: 50°C for 30 min.

   (b) Denature at 94°C for 2 min.

   (c) 30 cycles of PCR consisting of: 94°C for 15 s (denature), 55°C for 30 s (anneal), and 68°C for 1 min (extension).

5. Prepare reactions for electrophoresis by combining 10 μl PCR sample, 9 μl nuclease-free water, and 1 μl 10× DNA gel loading buffer.

6. Prepare DNA MW markers by combining 500 ng MW markers with 19 μl of water and 1 μl of DNA gel loading buffer.

7. Pre-run a 1.2% agarose E-Gel® at 75 Watts (constant) for 2 min using the power supply and E-Gel® base unit.

8. Turn off power supply, and load 20 μl of sample per lane.

9. In the lanes without sample, load 20 μl of water.

10. Turn power supply on and run at 75 Watts (constant) for 30 min.

11. Turn off power supply and remove gel from base unit.

12. Visualize the ethidium bromide stained gel by placing in the ultraviolet light box.

13. Both COX-1 and COX-2 PCR products should be a little over 1,800 base pair (bp) in size.

Table 1
PCR primer pairs

Primer designation	Primer DNA sequence[a]
COX1_forward	GACTGGATCCGCGCCATGAGCCGGAGTCTCTTGCTC
COX1_reverse	AGCTGCGGCCGCTCAGAGCTCTGTGGATGGTCGCTCC
COX2_forward	TAGCGGATCCGCTGCGATGCTCGCCCGCGCCCTG
COX2_reverse	AGTCGCGGCCGCCTACAGTTCAGTCGAACGTTC

[a]5′ to 3′.

3.1.4. Cloning of COX PCR Products into Plasmid Vector

In order to clone the PCR products into a plasmid vector, the products must first be purified, digested with restriction enzymes, and gel-purified for ligation to the vector. The vector must also be digested in order to accept the PCR products.

3.1.4.1. PCR Product Purification

1. To purify the PCR products, add 5 volumes of Buffer PB to 1 volume of the PCR sample and mix.
2. Place a QIAquick spin column in a 2 ml collection tube (provided as part of the kit).
3. Transfer PCR sample to the column and centrifuge for 1 min at $13,000 \times g$ at RT.
4. Discard flow-through and place the spin column back into the same tube.
5. Wash by adding 0.75 ml Buffer PE to the column and spin 1 min at $13,000 \times g$.
6. Discard flow-through and place the spin column into a new 1.5 ml tube.
7. Spin the column for an additional 1 min at $13,000 \times g$ to remove residual buffer.
8. Place spin column in a clean 1.5 ml centrifuge tube.
9. Elute DNA by adding 50 µl Buffer EB to the center of the column membrane.
10. Centrifuge the column for 1 min at $13,000 \times g$.

3.1.4.2. Restriction Enzyme Digestion

1. In order to digest each PCR product, combine 45.5 µl of PCR product with 5.5 µl React® 3 Buffer, 2 µl of BamHI, and 2 µl of NotI restriction enzymes.
2. Mix by pipetting 3–5 times, and incubate in 37°C water bath for 1.5 h.
3. Also prepare the insect cell expression vector, pFastBac1, for restriction digestion by combining 2 µg of vector (2 µl) with 2 µl React® 3 Buffer, 14 µl of water, and 1 µl each of BamHI and NotI restriction enzymes.
4. Mix by pipetting 3–5 times, and incubate in 37°C water bath for 1 h.
5. After 1 h at 37°C, add 1 unit of Shrimp Alkaline Phosphatase to pFastBac1 digestion, and continue incubation at 37°C for 30 min. This treatment will remove 5′ phosphate groups from the pFastBac1 fragments to prevent self-ligation.

3.1.4.3. Agarose Gel Electrophoresis and Gel Purification

1. To prepare the products for gel-purification, pre-run a 1.2% agarose E-Gel® at 75 Watts (constant) for 2 min using the power supply and E-Gel® base unit.
2. Prepare DNA MW markers by combining 500 ng MW markers with 19 µl of water and 1 µl of DNA gel loading buffer.

3. Load the restriction digestion reactions directly in a well of an E-Gel® without first adding sample buffer.

4. Load the MW markers, too, in another well.

5. In the lanes without sample, load 20 µl of water.

6. Turn power supply on and run at 75 Watts (constant) for 30 min.

7. Turn off power supply and remove gel from base unit.

8. Pry apart the E-Gel.

9. Place the ethidium bromide-stained agarose gel on the ultraviolet light box to visualize the ~4,800 bp pFastBac1 fragment and 1,800 bp COX PCR products.

10. Using disposable scalpels or a sharp knife, cut out the fragments and place into 450 µl Buffer QG. Agarose fragments typically weigh less than 100 mg, and the protocol recommends 3 volumes of Buffer QG for every 100 mg of agarose fragment, therefore, 450 µl is sufficient for this application.

11. Incubate in a 50°C water bath for 15 min to help dissolve the gel fragments, and mix by vortexing the tube every 3 min during the incubation to facilitate solubilization.

12. Place QIAquick spin columns in 2 ml collection tubes.

13. To bind DNA, apply the samples to the QIAquick columns, and centrifuge for 1 min at $13,000 \times g$ at RT.

14. Discard flow-through and place columns back in the same collection tubes.

15. To wash, add 0.75 ml of Buffer PE to spin columns and centrifuge for 1 min at $13,000 \times g$.

16. Discard the flow-through and transfer the spin columns to new centrifuge tubes.

17. Centrifuge the spin columns for an additional 1 min at $13,000 \times g$.

18. Place each spin column into a clean 1.5 ml centrifuge tube.

19. Elute the DNA using 50 µl of Buffer EB to the center of the QIAquick membrane and centrifuge the columns for 1 min at $13,000 \times g$.

20. Let the columns stand for 1 min, and then centrifuge for 1 min at $13,000 \times g$.

21. Store on ice or at 4°C until next step.

3.1.4.4. Ligation of PCR Products into Cloning Vector

1. Set up the ligations, below, using the Rapid DNA Ligation reagents:

   (a) COX-1/pFastBac1: 4 µl COX-1 PCR, 1 µl pFastBacI, 5 µl 2× T4 DNA

Ligation Buffer, and 0.5 μl T4 DNA Ligase.

    (b) COX-2/pFastBac1: 4 μl COX-2 PCR, 1 μl pFastBacI, 5 μl 2× T4 DNA

Ligation Buffer, and 0.5 μl T4 DNA Ligase.

    (c) pFastBac1: 4 μl water, 1 μl pFastBacI, 5 μl 2× T4 DNA

Ligation Buffer, and 0.5 μl T4 DNA Ligase.

2. Mix each by pipetting, and incubate at RT for 30 min.

3. During the incubation, thaw three aliquots of chemically competent cells on ice.

4. After incubation, add 1 μl of each ligation to an aliquot of cells.

5. Incubate cells on ice for 30 min.

6. Transfer cells to 42°C water bath for 30 s, and return to the ice bath for 2 min.

7. Add 250 μl TB and incubate at 37°C by shaking at 200–250 rpm for 1 h.

8. Place three LB carbenicillin plates at 37°C to warm up.

9. After the hour incubation, dump 5–10 glass beads onto each plate.

10. Spread 100 μl of cells onto each plate by shaking beads back and forth repeatedly.

11. Dump the beads, and incubate the plates at 37°C overnight.

12. The number of colonies on the plates containing the PCR products should outnumber the colonies on the pFastBac1 vector alone plate by at least 100-fold.

13. Pick and inoculate five colonies from each of the COX-1 and COX-2 plates into 4 ml of TB containing 100 μg/ml of carbenicillin using a sterile pipette tip. Use 15 ml tubes with loose-fitting caps.

14. Incubate cultures by shaking at 200–250 rpm at 37°C overnight.

3.1.5. Preparation and Sequencing of COX-1 and COX-2 Plasmids

1. Transfer 250 μl of culture for each clone into a 2 ml screw-capped tube for preparation of a seed stock.

2. Centrifuge the remaining overnight cultures at 5,000 × *g* for 5 min at RT.

3. During centrifugation, add 100 μl glycerol to the 250 μl cell aliquots, vortex thoroughly, and store at −80°C.

4. After centrifugation, discard all the liquid.

5. Re-suspend cell pellet in 250 μl Buffer P1, and transfer to a micro-centrifuge tube.

6. Add 250 μl Buffer P2 and mix thoroughly by inverting the tube 4–6 times.

7. Incubate 5 min at RT.

8. Add 350 µl Buffer N3 and mix thoroughly by inverting the tube 4–6 times. Shake vigorously if the material appears viscous and does not form a cloudy precipitate.

9. Centrifuge for 10 min at $13,000 \times g$ at RT.

10. Transfer the supernatants from to the spin column by pipetting.

11. Centrifuge for 60 s at $13,000 \times g$, and discard the flow-through.

12. Wash column by adding 0.75 ml Buffer PE and centrifuge 1 min at $13,000 \times g$.

13. Transfer the column to a new tube, and centrifuge for an additional 1 min at $13,000 \times g$ to remove residual wash buffer.

14. Place the QIAprep column in a clean 1.5 ml micro-centrifuge tube.

15. Elute the DNA by adding 50 µl Buffer EB to the center of each spin column, let stand for 1 min, and centrifuge for 1 min at $13,000 \times g$.

16. Estimate the DNA concentration by diluting 1 µl DNA into 99 µl water, and read the optical density (OD) at 260 nm and 280 nm.

17. Calculation: OD_{260} value $\times 100$ (dilution) $\times 50$ µg/ml = concentration in µg/ml.

18. Sequence each clone entirely to confirm.

3.2. COX-1 and COX-2 Protein Expression in Insect Cells

3.2.1. Generation of Recombinant Baculovirus Bacmids

1. For each construct, aliquot 20 µl of thawed DH10Bac cells into a 15 ml tube which has been chilled on ice.

2. Add 1 µl of each DNA from Subheading "Ligation of PCR Products into Cloning Vector" to the cells, mix gently, and incubate on ice for 30 min.

3. Heat-shock cells in a 42°C water bath for 45 s, cool on ice, add 200 µl SOC medium, and shake or rotate for 4 h at 37°C.

4. Add 800 µl of SOC, then make a 1:10 dilution of this in SOC; plate 100 µl of both of these on TGKIB plates (for 10% and 1%, respectively).

5. Incubate plates at 37°C for ~40 h, or until white colonies can be easily discriminated. Select well-separated, white colonies to start 5 ml overnights in LB with kanamycin, gentamicin, and tetracycline (see Note 1).

6. Use 1 ml of each overnight culture from step 1 of Subheading 3.2.1 for DMSO stock; spin remaining 4 ml to pellet cells.

7. Resuspend pellets in 300 µl solution P1 and transfer to 1.5 ml tube.

8. Add 300 µl solution P2, invert gently 4–6 times to mix, and incubate up to 5 min at RT.

9. Add 300 µl solution P3, invert gently 4–6 times, and set on ice for 5–10 min.

10. Spin top speed in microfuge 10 min, decant into clean tube, add 800 µl isopropanol, and incubate 10 min on ice.

11. Spin 15 min at top speed in microfuge, wash pellet in 500 µl 70% EtOH, and spin again for 5 min.

12. Remove EtOH carefully (pellets may be loose), and resuspend pellets in 40 µl TE (see Note 2).

13. Screen for insert using PCR with M13 primers and 1 µl of the miniprep DNA.

3.2.2. Transfection of Bacmids into Insect Cells

1. In a six-well tissue culture plate, seed 9×10^5 Sf9 cells per well in 2 ml Sf-900 II SFM; allow cells to attach at 27°C for at least 1 h.

2. Prepare the following solution in sterile 12×75 tubes, one tube per DNA sample: 200 µl Sf-900 II SFM, 5 µl Cellfectin Reagent, and 5 µl of the appropriate miniprep DNA; mix gently and incubate 15–45 min at RT.

3. Wash the cells once with 2 ml Sf-900 II without antibiotics, just prior to transfection.

4. Add 1 ml Sf-900 II to each tube containing the DNA-Cellfectin mixtures. Mix gently, add carefully to washed cells, and incubate cells ~5 h at 27°C.

5. Remove the transfection mixtures, feed the cells 3 ml Sf-900 II containing 5% FBS, and incubate 3 days at 27°C.

6. Harvest the medium which now contains recombinant virus, spin out dead cells and debris, and store the supernatant at 4°C; feed the wells 3 ml fresh complete medium and incubate for 3 more days.

7. Harvest as before, and use this second harvest to generate larger virus stocks.

3.2.3. Baculovirus Stock Generation

1. Seed 250 ml Sf9 cells at 1×10^6/ml in a disposable Erlenmeyer shake flask in Sf-900 II medium containing 5% FBS, for each virus.

2. Add 250 µl of the second medium harvest from step 3 of Subheading 3.2.1 containing the desired recombinant viruses to the cells.

3. Monitor viability and cell diameter by Cedex. Cells should stop growing after ~24 h, and diameters should increase at least 4–5 µm (see Note 3).

4. Harvest medium when the average cell diameter has increased and the viability is between ~50% and 70%; store at 4°C. This is called P1 virus stock, and usually takes between 3 and 5 days.

5. To titer virus, seed Sf9 cells in 6-well dishes at 0.9×10^6 cells/well in SF900-II medium; allow attaching for at least 1 h at 27°C. Cells should not be confluent since they will grow over the course of the assay.

6. Prepare dilutions of each P1 virus stock to be titered, in culture medium. Generally, dilutions of 10^{-6}, 10^{-7}, and 10^{-8} will give plaques in a good number to be counted.

7. Once cells have plated, remove medium and add 1 ml virus dilution per well, in duplicate; incubate 1–2 h with occasional gentle rocking.

8. During virus incubation, prepare sufficient Agarose Plaquing Medium for 3 ml per well; microwave agarose prior to mixing, and keep it liquefied in a beaker of warm water (~45–50°C).

9. At the end of the incubation, remove the virus by aspiration, and slowly add the warmed plaquing medium evenly to the wells. Allow to solidify in the wells before moving plates to a 27°C incubator.

10. Incubate for 4–6 days, until plaques are well formed, then stain with 1 ml/well MTT dye, allowing several hours for the color to develop. Plaques will appear clear in a dark purple background.

11. Count plaques, calculate titer: Number of Plaques/Dilution of Virus = PFU/ml.

3.2.4. Optimization of Insect Cell Protein Expression

1. Inoculate 3 small shake flasks with 1×10^6 Sf9 cells/ml in Sf-900 II medium, and infect at time $T = 0$ with recombinant virus at multiplicities of infection (MOI) of 0.2, 1.0 and 5.0, using the titer determined in step 5 of Subheading 3.2.1.

2. Harvest samples for Western analysis and/or activity determination from each of the flasks at $T = 24$, 48, and 72 h.

3. Determine the optimal virus inoculum and time of harvest from the results obtained with the small-scale samples; use these conditions to produce protein at larger scales as needed.

3.3. Results

The coding regions for human COX-1 and COX-2 are easily cloned from LPS-stimulated U937 cells via PCR, and the resulting PCR products are sub-cloned into the baculovirus recombinant vector, pFastBac1. Then, bacmid plasmids are derived from the pFastBac1 plasmids to transfect insect cells for generation of viral stocks. Insect cells are infected at an MOI that maximizes the production of COX enzyme.

In baculovirus-infected insect cells human COX-1 is typically produced at 5–10% of the levels of that of human COX-2. In this case, COX-1 is expressed at 0.3 µg/ml while COX-2 is expressed at 6.3 µg/ml at the 10 L scale.

4. Notes

In all cases using pre-manufactured kits, the protocols provided by the manufacturers were followed except when the RNA or DNA mass was thought to be low (<1 µg). For example:

1. LPS-stimulation of U937 cells is required for up-regulation of the COX-2 message while COX-1 message levels remain relatively unchanged.

2. We recommend modifying the column purification protocols to improve yields by spinning columns at low speeds ($\leq 500 \times g$) for 1 min after loading the sample onto the column followed by the standard high-speed centrifugation. This low-speed centrifugation step improves retention of nucleic acids on the column.

3. We also recommend pre-heating elution buffers to 65 °C for DNA purification protocols to maximize the elution steps. High-temperature elution of RNA is not recommended due to RNA instability.

References

1. Shimokawa T, Smith WL (1992) Expression of prostaglandin endoperoxide synthase-1 in a baculovirus system. Biochem Biophys Res Commun 183(3):975–982

2. Miller DB, Munster D, Wasvary JS, Simke JP, Peppard JV, Bowen BR, Marshall PJ (1994) The heterologous expression and characterization of human prostaglandin G/H synthase-2 (COX-2). Biochem Biophys Res Commun 201(1):356–362

3. Barnett J, Chow J, Ives D, Chiou M, Mackenzie R, Osen E, Nguyen B, Tsing S, Bach C, Freire J, Chan J, Sigal E, Ramesha C (1994) Purification, characterization and selective inhibition of human prostaglandin G/H synthase 1 and 2 expressed in the baculovirus system. Biochim Biophys Acta 1209(1):130–139

4. Gierse JK, Hauser SD, Creely DP, Koboldt C, Rangwala SH, Isakson PC, Seibert K (1995) Expression and selective inhibition of the constitutive and inducible forms of human cyclooxygenase. Biochem J 305(Pt. 2):479–484

5. George HJ, Van Dyk DE, Straney RA, Trzaskos JM, Copeland RA (1996) Expression purification and characterization of recombinant human inducible prostaglandin G/H synthase from baculovirus-infected insect cells. Protein Expr Purif 7(1):19–26

6. Gierse JK, Staten NR, Casperson GF, Koboldt CM, Trigg JS, Reitz BA, Pierce JL, Seibert K (2002) Cloning, expression, and selective inhibition of canine cyclooxygenase-1 and cyclooxygenase-2. Vet Ther 3(3):270–280

7. Zhang WY, Yang XN, Jin DZ, Zhu XZ (2004) Expression and enzyme activity determination of human cyclooxygenase-1 and -2 in a baculovirus-insect cell system. Acta Pharmacol Sin 25(8):1000–1006

Chapter 5

Expression of Cyclooxygenase Isoforms in the Baculovirus Expression System

John C. Hunter, Natalie M. Myres, and Daniel L. Simmons

Abstract

Prostaglandin synthase-1 and 2, also known as cyclooxygenase-1 and 2 (COX) catalyze the rate limiting step in the conversion of arachidonic acid to prostaglandins and other potent lipid mediators. As such, COX enzymes represent important pharmacological targets in the treatment of pain, inflammation, and fever. However, due to the typically low expression level of COX enzymes in most primary tissues, large amounts of highly purified enzyme for in depth study is difficult to obtain. Overexpression of COX in the baculovirus expression system represents a quick and efficient way to overcome this problem and obtain large amounts of catalytically active enzyme. Here, we present a step-by-step approach for COX expression in the baculovirus system.

Key words: Cyclooxygenase isoforms, COX expression, Baculovirus expression, BaculoDirect

1. Introduction

Baculovirus expression is a widely used approach for recombinant protein generation that has been utilized for over 20 years (1). Baculovirus-infected insect cell expression has many advantages over other commonly used mammalian expression systems; namely, a very high level of protein production, an ability to coexpress a number of large genes and nearly equivalent posttranslational modification of proteins (2). For these reasons, baculovirus expression has been widely used in the expression study of COX genes. COX isoforms expressed in insect cells are catalytically active and can be purified to near homogeneity (3).

Samir S. Ayoub et al. (eds.), *Cyclooxygenases: Methods and Protocols*, Methods in Molecular Biology, vol. 644, DOI 10.1007/978-1-59745-364-6_5, © Springer Science+Business Media, LLC 2010

2. Materials

2.1. Initiation and Maintenance of Sf9 Cells

1. Sf9 frozen cells (Included in BaculoDirect kit, Invitrogen, Carlsbad, CA).

2. Grace's insect media with lactalbumin hydrolysate, yeasto-late & gentamicin (Lonza, Walkersville, MD) supplemented with 10% (vol/vol) fetal bovine serum (Sigma-Aldrich, St. Louis, MO).

3. 0.4% trypan blue (modified) in phosphate buffered saline (MP Biomediacls, LLC, Solon, OH).

4. Culture plates and roller bottles (Sarstedt, Germany).

2.2. Cloning of COX Isoforms into Baculovirus Vectors

1. pENTR/D-TOPO® Cloning Kit (Invitrogen, Carlsbad, CA).

2. GeneElute Mini Prep Kit (Sigma-Aldrich, St. Louis, MO).

3. BaculoDirect™ C-Term Expression Kit (Invitrogen, Carlsbad, CA).

4. LB Broth: 10 g Bacto-tryptone, 5 g yeast extract, 10 g NaCl, ddH$_2$O up to 1,000 mL, sterilized by autoclaving. Add kanamycin once cool.

5. LB – agar plates: 10 g Bacto-tryptone, 5 g yeast extract, 10 g NaCl, 15 g Agar, ddH$_2$O up to 1,000 mL, sterilized by autoclaving. Add kanamycin once cooled to ~50°C and pour into 100 cm^2 plates.

6. Kanamycin (Sigma-Aldrich, St. Louis, MO).

7. TE buffer: 10 mM Tris, 1 mM EDTA, pH 8.0.

2.3. Plaque Assay of Virus Stock

1. 5-bromo-4-chloro-3-indolyl-beta galactoside (X-gal) (Sigma-Aldrich, St. Louis, MO).

2. Ganciclovir (Included in BaculoDirect kit, Invitrogen, Carlsbad, CA).

3. Methods

3.1. Initiation and Maintenance of Sf9 Cells

1. Initiate adherent cell cultures by quickly thawing a tube of Sf9 cells. This can be done by warming it in your hand with gentle mixing.

2. While working in a sterile cell culture hood, pipette out the thawed aliquot of cells and transfer them to a 25 cm^2 culture flask containing 20 mL Grace's insect media with 10% FBS warmed to room temperature (see Note 1).

3. Distribute the cells evenly by briefly rocking the flask and incubate at room temperature for 60 min to allow the cells to adhere.

View the cells under a microscope to verify that they are attached to the plate.

4. Aspirate off the media once the cells are attached and add 20 mL of fresh media and place in an incubator set to 27°C.

5. Once the cells have reached 90–100% confluence (48–72 h), dislodge the cells from the flask by running media over the bottom of the flask a few times. Dilute the cells one third and divide into three new flasks.

6. Suspensions cultures may be initiated from adherent cultures by harvesting adherent cells and adding these to large roller bottles with media at a concentration of 1×10^6 cells/mL.

7. Cells should be passaged by diluting 1:2 in media once the cell density has reached 2×10^6 cells/mL. Suspension cultures should double approximately every 24 h.

8. Cell density and viability is measured by hemocytometer and trypan blue assay.

9. The trypan blue assay is performed by removing 20 µL of cells and mixing with 20 µL 0.4% trypan blue. Ten microliters of the cell suspension-trypan blue mixture is placed on a hemocytometer and the cells counted within 60 s of trypan blue addition. Nonviable cells will be stained blue (see Note 2).

3.2. Cloning of COX Isoforms into Baculovirus Vectors

The coding region of the desired cDNA is cloned into the cloning site of the baculovirus vector. In the past, once cloning of the cDNA into a Baculovirus vector was achieved, this was followed by a rather laborious process of performing homologous recombination in the insect cell host and isolating recombinant virus by repeated plaque assays. However, now a direct recombination-based vector BaculoDirect (Invitrogen) allows cloning into the host to be done recombinantly in vitro in a 1-h reaction. Moreover, this new method minimizes the need for extensive plaque purification.

The specific protocol (4) for the BaculoDirect system is supplied with the kit from Invitrogen and should be referenced for any modifications by the manufacturer. The procedure to be followed is outlined here (see Note 3).

1. The desired cDNA fragment is amplified by PCR using a proofreading enzyme such as platinum Pfx (Invitrogen) and gel purified. When designing primers for amplification, it is necessary to add four nucleotides, CACC to the 5′ end of the forward primer. This allows for directional cloning into the pENTR/D-TOPO® vector (Invitrogen). We suggest using this transfer vector as it allows for rapid directional ligation. (As with the BaculoDirect system the specific protocol (5) for ligation is included in the Gateway Entry kit and should also be referenced for optimal efficiency.)

2. Once the cDNA has been purified, measure the optical density to determine the concentration of cDNA. Gently mix

15–20 ng of cDNA in 4 µL of molecular biology grade water with 1 µl salt solution and 1 µL of TOPO vector and incubate at room temperature for 5 min.

3. Add 2 µL of the reaction to a tube of thawed One Shot® chemically competent *E. coli* (Invitrogen) and incubate on ice for 5 min.

4. Heat shock the cells for 30 s at 42°C and place back on ice. Add 250 µL of prewarmed SOC media and incubate at 37°C with shaking for 1 h.

5. Spread 150 µL aliquots onto two prewarmed agar plates containing 50 µg/mL Kanamycin and incubate at 37°C overnight. The next day pick 10 or more colonies and PCR screen for ligation and correct orientation of your cDNA insert. Select three or four colonies which have the insert in the correct orientation.

6. Inoculate 3 mL of LB Broth containing 50 µg/mL Kanamycin with the three or four best colonies and incubate at 37°C overnight. The next day, purify the plasmids using a GeneElute mini prep kit from Sigma. We recommend dideoxy sequencing each of the vectors to verify the cDNA sequence and orientation of the cDNA insert is correct. Select one of the vectors to use for generation of the Baculovirus.

7. Ethanol precipitate the vector and resuspend into TE buffer pH 8.0 at a concentration of ~100 ng/mL.

8. Mix 10 µL of the provided BaculoDirect linear DNA with 1–2 µL of purified entry vector. Add TE buffer up to 16 µL.

9. Thaw the LR Clonase™ II enzyme mix on ice and mix gently.

10. Add 4 µL of the enzyme mix to your ligation reaction to achieve a total volume of 20 µL and flick the tube to mix gently.

11. Incubate the tube at 25°C for 1 h.

12. The kit from Invitrogen also contains a positive control pENTR vector which should be used in a separate cloning reaction to ensure that each step proceeds as expected.

    If done correctly, this process should give >95% recombination. We recommend you verify the ligation at this step by polymerase chain reaction (PCR) amplification of a very small amount of your product using the provided polyhedrin forward primer and the V5 reverse primer or other primers of your own design. The PCR product is then run on a 1% agarose gel and stained with ethidium bromide to visualize.

3.3. Transfection of Sf9 Cells

1. Once the viral DNA has been constructed, Sf9 cells are transfected to produce infectious virions.

2. Plate ~8×10^5 Sf9 cells onto one of the wells on a 6-well plate with 2 mL supplemented Grace's insect media. Incubate the

plate at 27°C for 1 h to allow the cells to adhere to the bottom of the well.

3. In the mean time, take 5–10 µL of your ligation reaction from step 7 in Subheading 3.2 and mix with 100 µL of Grace's insect media in a 1.5 mL eppendorf tube. Make sure to use unsupplemented media as the proteins and other components in the supplemented media may interfere with the transfection.

4. In a separate tube, combine 6 µL of well mixed Cellfectin® Reagent (Provided with the BaculoDirect kit) with 100 µL of unsupplemented Grace's insect media.

5. Combine the two solutions and gently mix by inversion. Let this sit at room temperature for 15–45 min. After incubating, add another 800 µL of unsupplemented media and gently mix again.

6. Inspect the cells in the 6-well plate using a microscope to ensure that the cells have adhered to the surface. Then carefully aspirate the media off the cells.

7. Wash the cells with 2 mL of unsupplemented media. Add the solution from step 5 drop-wise to the well without disturbing the cell layer. Incubate the cells at 27°C for 5 h.

8. After the 5 h incubation, aspirate off the solution and add 2 mL of supplemented Grace's insect media containing 100 µM ganciclovir to the well. Incubate at 27°C for 96 h. Ganciclovir is a nucleoside analog which is activated by herpes simplex virus type 1 thymidine kinase (HSV1tk). Upon activation it is incorporated into DNA to inhibit replication. The HSV1tk gene is encoded on the BaculoDirect linear baculovirus DNA and is removed upon recombination with the Gateway vector. Thus, ganciclovir allows you to select against nonrecombinant virus.

9. After the 96-h incubation, look at the cells under the microscope to determine if they are showing signs of viral infection. Some of the signs to look for include, reduced cell proliferation, enlargement of the cell relative to noninfected cells, and reduced cell viability as measured by the trypan blue assay described earlier. It may be necessary to let the cells incubate longer than the 96 h for optimal viral expression.

10. Remove the media from the well and transfer it to a 15 mL conical vial. Centrifuge at 4,000 × g for 5 min to pellet cells and any other debris. The supernatant is your P1 viral stock.

3.4. Plaque Assay of Virus Stock

A plaque assay should be performed in order to assess the efficiency of the recombination reaction or to generate a virus stock from a single clone. The plaque assay takes advantage of the fact that nonrecombinant virus contains the lacZ gene encoding β-galactosidase. Upon incubation with 5-bromo-4-chloro-3-indolyl-beta galactoside (X-gal), β-galactosidase cleaves X-gal to form 5-bromo-4-chloro-3-hydroxyindole which is oxidized to an insoluble blue precipitate. Thus, blue colonies indicate the presence of nonrecombinant virus. The plaque assay is performed as follows.

1. Three aliquots of 5×10^6 Sf9 cells in 5 mL supplemented Grace's insect media with 100 µM ganciclovir are plated out on three 100 mm tissue culture plates and incubated at 27°C for 30–45 min to allow the cells to adhere. (View the cells under a microscope to verify.)

2. After the cells have attached, remove all but 2 mL of the media and slowly add 1 mL of the P1 virus stock diluted 1:100, 1:1,000, and 1:10,000 in supplemented Grace's insect medium with ganciclovir. Incubate at room temperature for 1 h with rocking.

3. Prepare an agarose overlay mixture by combining 10 mL of a 2.5% agarose solution (melted) with 10 mL complete Grace's insect media. Keep this in a 47°C water bath. Immediately before use, take 5 mL of this and mix with 5 mL of media containing 150 µg/mL X-gal.

4. After the 1-h incubation in step 2, remove the media from the cell layer and carefully overlay the cells by pipetting 10 mL of the agarose solution from step 3 over the cells. Be sure not to disturb the monolayer.

5. Incubate over-laid cells for 5–7 days at 27°C. Inspect daily for the appearance of blue plaques indicating the presence of nonrecombinant virus.

6. If your plaque assay demonstrated that your virus is free from nonrecombinants, it is acceptable to simply amplify your P1 viral stock from Subheading 3.3 to generate a higher titer stock for protein expression (Subheading 3.5) without selecting plaques and amplifying these. However, if blue plaques are observed, generate a viral stock from one of the nonblue, recombinant plaques as described below.

7. Using a sterile glass Pasteur pipette, poke out a chosen plaque from the overlay. Add this agarose plug to a well of a 24-well plate that has been preseeded with 2.5×10^5 cells per well in 1 mL of complete Grace's media.

8. Incubate the cells at 27°C for 96 h and proceed to harvest viral P1 stock as done in steps 9 and 10 of Subheading 3.3. Use this P1 stock for the rest of the protocol.

3.5. Generation of High Titer Virus for Protein Expression

1. For most protein expression work, it will be desirable to have a large amount (500 mL) of high titer virus stock on hand (see Note 4). Amplification of the stock is achieved by seeding 75 cm² culture plates with 2×10^6 log phase cells in

20 mL. (Incubate the cells long enough to allow the cells to adhere.) 20 µL of each P1 stock to be amplified is added to the plate. These are then incubated at 27°C for ~10 days.

2. Following the incubation, the cells and media are harvested and transferred to sterile 15 mL conical vials. These are then centrifuged at $4,000 \times g$ for 5 min and the supernatant recovered to yield P2 viral stock.

3. Add 5 mL of each P2 viral stock to be amplified to a 500 mL suspension culture in roller bottles containing 1×10^9 Sf9 cells in supplemented Grace's insect media with ganciclovir.

4. Incubate the cells at 27°C for 10–15 days. Monitor the level of cell lysis using the trypan blue exclusion assay described in Subheading 3.1. When cells reach 30% viability, harvest the cells and centrifuge at $4,000 \times g$. Collect the supernatant as the P3 viral stock.

5. At this point, the titer of the P3 virus stock should be assessed to determine if it is sufficient for protein expression studies. This is done by repeating the plaque assay described in Subheading 3.4. Because of the amplified titer of the P3 stock compared to the P1 stock, 10^{-6}, 10^{-7}, and 10^{-8} dilutions of the P3 stock should be used for the assay.

6. Following the 5–7 day incubation, the number of plaques formed is counted. The viral titer is calculated based on the following formula:

$$\text{Plaque Forming Units (pfu)} / \text{mL} = \frac{(\# \text{ of Plaques})}{(\text{Dilution of P3 stock})}$$

3.6. Protein Expression

1. In order to assess the level of protein expression, high titer P3 viral stock is used to infect Sf9 cells and the proteins recovered and assayed by SDS-PAGE and immunoblot (Fig. 1).

2. 1×10^6 cells/well are plated on a six well plate with 1.5 mL of supplemented Grace's insect media with ganciclovir and are allowed to adhere for 1 h. The amount of P3 virus stock to add to each well is calculated based upon the desired multiplicity of infection (MOI) as shown below. Typically, cells are infected at a MOI between 5 and 10 for expression experiments.

$$\text{Volume of Viral Stock to add} = \frac{(\backslash \# \text{ of Cells}) \times (\text{MOI})}{(\text{Virus Titer})}$$

$$\text{Example of Calculation: } 50 \backslash \mu L = \frac{(1 \times 10^6 \text{ cells}) \times (5 \text{ pfu} / \text{cell})}{(1 \times 10^8 \text{ pfu} / \text{mL})}$$

3. The calculated amount of P3 viral stock is added to each well. Cells are harvested by gently scraping them off the plate with

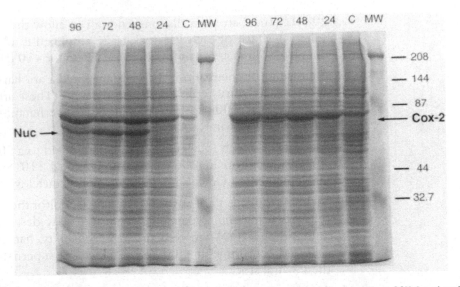

Fig. 1. *A representative coomassie blue stained gel showing baculovirus expressed recombinant COX-2 and nucleobindin (Nuc). Sf9 cells were infected at an MOI of 8 with recombinant COX-2 and Nuc high titer stocks. Aliquots of the infected cultures were removed at 24-h intervals postinfection. The resulting cell lysates were subjected to electrophoresis on a 10.5% SDS-PAGE gel and analyzed for protein expression with coomassie blue stain. Arrows indicate the protein bands corresponding to Nuc and COX-2 respectively. (From ref. 6).*

a sterile cell scraper after 24–96-h incubation at 27°C. The optimal incubation time for maximum protein recovery is specific to that protein and should be determined by a time course assay. It is also important to note that cell viability varies based upon the MOI used. Therefore, the MOI should be kept constant between experiments and cell viability measured between samples for consistency.

4. The harvested cells are transferred to a 15 mL conical tube and cell viability is assayed by the trypan blue exclusion assay. Cells are then pelleted by centrifugation at $500 \times g$ for 5 min (Fig. 2b).

5. The supernatant is aspirated off and the cell pellets are resuspended and analyzed by SDS-PAGE and immunoblot for protein expression (Fig. 2a).

4. Notes

1. It is important to perform all steps in a sterile culture hood to avoid bacterial or fungal contamination of the insect cell cultures. Be sure to use sterile equipment, pipettes, etc. and to change gloves often.

a

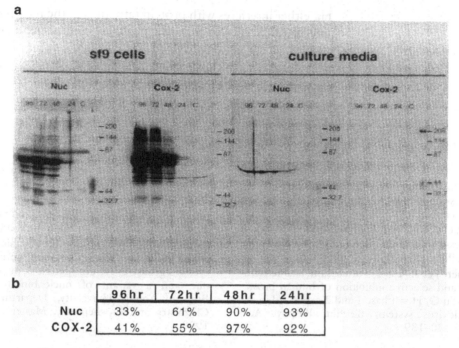

b

	96hr	72hr	48hr	24hr
Nuc	33%	61%	90%	93%
COX-2	41%	55%	97%	92%

Fig. 2. *A representative autoradiogram of an immunoblot of recombinant COX-2 and Nuc.* (**a**) Aliquots of sf9 cells expressing recombinant COX-2 or Nuc were isolated at 24-h intervals (24–96) post-infection. The resulting cell lysates and culture media were immunoblotted using anti-COX and anti-Nuc respectively. (**b**) At each 24-h interval, the corresponding percentage of viable cells was measured by trypan blue exclusion.

2. When performing a trypan blue viability assay, it is important to count the number of viable cells within 60 s of trypan blue addition. Trypan blue is toxic to the cells, and a longer incubation will lead to an artificially low viability count.

3. Baculovirus constructs allow the production of proteins which may be tagged or expressed as fusion proteins (e.g. GST fusion) depending on the expression vector used and whether a stop codon is included at the end of the cDNA open reading frame that is cloned into the vector. Vectors for these uses are marketed by Invitrogen. With regard to tagging of COX proteins for purification, his-tags are most useful for the production of COX fragments. Commercial his-tag vectors will place the his-tag either on the N or C terminus of the COX fragment depending on the vector used. However, if rapid purification of enzymatically active, his-tagged COX proteins is desired, the poly-histidine tag should be inserted internally at the C-terminal end of the COX N-terminal signal peptide prior to cloning the cDNA into the expression vector. Otherwise, poor expression is typically achieved, because a C-terminal his-tag interferes with COXs' carboxy-terminal ER-retention signal or otherwise destabilizes the protein and an N-terminal

tag either interferes with protein targeting to the lumen of the ER or will be cleaved from the protein by signal peptidase.

4. Viral stock solutions may be stored for long periods of time at 4°C protected from light. For long term storage, an aliquot may be frozen at –80°C. Repeated freeze thaw cycles should be avoided.

References

1. Kost TA, Condreay JP, Jarvis DL (2005) Baculovirus as versatile vectors for protein expression in insect and mammalian cells. Nat Biotechnol 23:567–575

2. Hu YC (2005) Baculovirus as a highly efficient expression vector in insect and mammalian cells. Acta Pharmacol Sin 26(4):405–416

3. Barnett J et al (1994) Purification, characterization and selective inhibition of human prostaglandin G/H synthase 1 and 2 expressed in the baculovirus system. Biochim Biophys Acta 1209:130–139

4. BaculoDirect™ Baculovirus expression systems user manual, Invitrogen, Version H, 16 June 2006

5. pENTR™ Directional TOPO® Cloning kits user manual, Invitrogen, Version G, 6 April 2006

6. Myres NM (1999) Development of high level eukaryotic expression systems for synthesis of cyclooxygenase 1, cyclooxygenase 2, nucleobindin, and a variant of nucleobindin (N4), Brigham Young University, Department of Chemistry and Biochemistry, Master's thesis, 1999

Chapter 6

Peroxidase Active Site Activity Assay

Kelsey C. Duggan, Joel Musee, and Lawrence J. Marnett

Abstract

Cyclooxygenase enzymes house spatially distinct cyclooxygenase- and peroxidase-active sites. The two-electron reduction of peroxides to their corresponding alcohols by the heme bound in the peroxidase-active site converts the heme to a ferryloxoprotoporyphrin cation radical, with a reductant providing the two electrons necessary to bring the heme back to its resting state. The ferryloxoprotoporyphrin cation radical can abstract a hydrogen atom from a tyrosine residue in the cyclooxygenase-active site, activating the oxygenase functionality. The tyrosyl radical subsequently abstracts a hydrogen atom from the cyclooxygenase substrate, arachidonic acid, leading to its oxygenation and the formation of a hydroperoxy endoperoxide intermediate, PGG_2. The peroxidase functionality reduces PGG_2 to the hydroxy endoperoxide, PGH_2, which serves as the precursor to downstream prostaglandins and thromboxane. The peroxidase activity of cyclooxygenase enzymes can be assayed by quantifying the oxidation of a peroxidase reductant or the reduction of a hydroperoxide substrate. Here we describe a spectrophotometric assay used to measure the oxidation of a reductant, 2,2'-azino-*bis* (3-ethylbenzthiazoline-6-sulfonic acid) (ABTS), as well as a high-performance liquid chromatography method for the measurement of the conversion of 5-phenyl-4-pentyl hydroperoxide (PPHP) to its corresponding alcohol. The first provides a continuous but indirect assay of peroxidase activity, whereas the second provides a discontinuous but direct assay.

Key words: Cyclooxygenase, PGHS, Peroxidase, ABTS, PPHP, Peroxidase reductant

1. Introduction

The duality of function of the cyclooxygenase enzymes, with spatially distinct peroxidase- and oxygenase-active sites, is an elegant solution to the control of prostaglandin synthesis (1). The oxygenation of arachidonic acid to form the hydroperoxy endoperoxide, PGG_2, occurs at the cyclooxygenase-active site (2). PGG_2 then undergoes a rapid two-electron reduction at the peroxidase-active site resulting in the formation of the hydroxy endoperoxide, PGH_2 (3). These two active sites are functionally coupled as the

Samir S. Ayoub et al. (eds.), *Cyclooxygenases: Methods and Protocols*, Methods in Molecular Biology, vol. 644,
DOI 10.1007/978-1-59745-364-6_6, © Springer Science+Business Media, LLC 2010

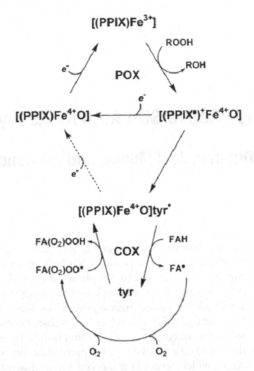

Fig. 1. The Branched Chanin Mechansim. Functional coupling of the peroxidase- and cyclooxygenase-active sites. (Reproduced with permission of American Chemical Society).

peroxidase site is required to activate the oxygenase functionality. This phenomenon has been best explained by the branched chain mechanism (Fig. 1) (4, 5).

The catalytic mechanism of the peroxidase site is similar to that of other heme peroxidases (1). Briefly, the two-electron reduction of the hydroperoxide substrate to the corresponding alcohol occurs in conjunction with the two-electron oxidation of heme $[(PPIX)Fe^{3+}]$ to form a ferryloxoprotoporyphrin cation radical $[(PPIX)^{+}Fe^{4+}O]$. Two subsequent one-electron reductions bring the enzyme back to its resting state $[(PPIX)Fe^{3+}]$. The peroxidase reductant, also called the reducing cosubstrate, provides the electrons that are necessary for reduction (1). Therefore, measuring the reduction of the hydroperoxide substrate or the oxidation of the peroxidase reductant can be useful for monitoring the peroxidase reaction (1). As per the aforementioned branched mechanism (Fig. 1), the ferryloxoprotoporyphrin cation radical $[(PPIX)^{+}Fe^{4+}O]$ can abstract a hydrogen atom from a tyrosine residue at the oxygenase-active site. This tyrosyl radical subsequently abstracts a hydrogen atom from arachidonic acid and similar substrates (FAH) (Fig. 1). This is the point of coupling between the two active sites, facilitating the introduction of two molecules of oxygen into the carbon framework of FAH. The regeneration of the tyrosyl radical at the end of oxygenation of

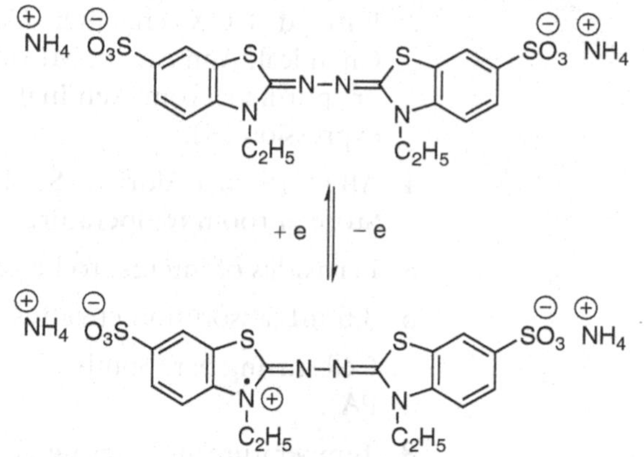

Fig. 2. Oxidation of 2,2′-azino-*bis*(3-ethylbenzthiazoline-6-sulfonic Acid).

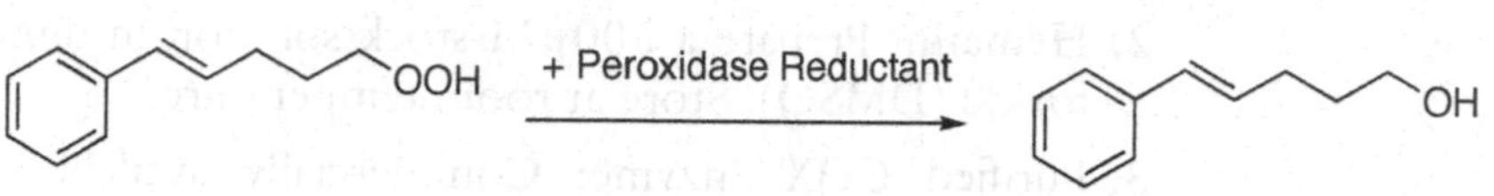

Fig. 3. Reduction of 5-phenyl-4-pentenyl hydroperoxide to the product alcohol.

FAH explains the expendability of the peroxidase site after activation of the oxygenase functionality.

The oxidation of 2,2′-azino-*bis*(3-ethylbenzthiazoline-6-sulfonic Acid) (ABTS), which serves as a peroxidase reductant, can be easily monitored spectrophotometrically. The radical produced upon the oxidation of ABTS (ABTS$^+$) (Fig. 2) is brilliantly green-blue and has a λ_{max} of ~414 nm (6). Since the ABTS reaction is continuous, it is ideal for the determination of peroxidase kinetics. As an alternative method for the quantification of peroxidase activity, the reduction of the synthetic hydroperoxide, 5-phenyl-4-pentenyl hydroperoxide (PPHP) to the resulting alcohol, 5-phenyl-4-pentenyl-alcohol (PPA) in the presence of reducing substrates (Fig. 3) may be measured (7). From incubation mixtures, PPHP and PPA can be easily separated using high-performance liquid chromatography and quantified by ultraviolet detection ($\lambda_{max} = 254$ nM) (7).

2. Materials

2.1. ABTS Assay

1. Hematin: Prepare a 5 µM stock solution in dimethyl sulfoxide (DMSO). Store at room temperature.

2. Reaction buffer: 80 mM Tris–HCL (pH 8.0). Store at room temperature.

3. Purified COX enzyme: Commercially available (Cayman Chemical, Ann Arbor, MI or Sigma-Aldrich, St. Louis, MO), or purified from seminal vesicles or from recombinant expression (8).

4. ABTS (Sigma-Aldrich, St. Louis, MO): Protect from light. Store at room temperature.

5. Peroxides of interest to be tested.

6. 3.5 mL absorption cuvettes.

7. Cell stirring bar (Spinbar®, VWR International, West Chester, PA).

8. Temperature and stirring control-fitted spectrophotometer.

2.2. PPHP Assay

1. Reaction buffer: 100 mM Tris–HCl (pH 8.0). Store at room temperature (see Note 1).

2. Hematin: Prepare a 500 μM stock solution in dimethyl sulfoxide (DMSO). Store at room temperature.

3. Purified COX enzyme: Commercially available (Cayman Chemical, Ann Arbor, MI or Sigma-Aldrich, St. Louis, MO), or purified from seminal vesicles or from recombinant expression (8).

4. PPHP (Cayman Chemicals, Ann Arbor, MI): Provided as a stock solution (5 mg/mL) in ethanol. Store long term at –20°C.

5. Quench solution: Ethyl acetate with 0.5% acetic acid. Store at room temperature. Place on ice before use in assay.

6. Reducing cosubstrates of interest.

7. Inhibitors of interest.

8. HPLC grade methanol.

9. HPLC grade water.

10. Luna C$_8$ 3 μ HPLC column (150 × 2.00 mm) (Phenomenex, Torrance, California).

3. Methods

ABTS Assay: The ABTS assay can be utilized to determine the peroxidase activity of COX enzymes by measuring the oxidation of the reducing cosubstrate (ABTS) to its corresponding cation radical (ABTS$^{+\cdot}$). With slight modifications, the assay can also be used to examine the ability of a compound of interest to act as a peroxidase substrate. This continuous spectrophotometric assay is also ideal to monitor reaction kinetics. The reaction volume of the protocol described below will total 2 mL but may be modified as the investigator sees fit.

PPHP assay: The PPHP assay can be utilized to determine the peroxidase activity of COX enzymes by measuring the reduction of the hydroperoxide substrate (PPHP) to its corresponding alcohol (see Note 2). With slight modifications, the assay can also be used to examine the ability of a compound of interest to act as a peroxidase reductant or inhibitor of peroxidase activity. The reaction volume of the protocol described below will total 500 μL but may be modified as the investigator sees fit.

3.1. ABTS Assay

1. Prepare a 100 mM ABTS solution in 80 mM Tris–HCL buffer and keep on ice. Prepare enough solution to provide 20 μL for each 2.0 mL reaction. The reduced form is colorless to light yellow green, while the oxidized form is a brilliant green-blue (see Note 3).

2. Prepare stock solutions of the peroxide substrates (see Notes 4 and 5). Water-soluble peroxides, such as hydrogen peroxide should be dissolved in the reaction buffer. Other stocks may be prepared in water-miscible solvents such as ethanol or DMSO. Peroxide stocks should be made at 40-fold the desired final concentration, ensuring an addition of 50 μL into the total 2 mL reaction that leads to the appropriate dilution.

3. Prepare a master reaction mix containing 1974 μl of reaction buffer and 20 μl of 100 mM ABTS per reaction.

4. For each reaction, aliquot 1994 μL of the master mix to a borosilicate tube. ABTS is labile upon exposure to light and air; therefore, cover the tubes with aluminum foil (see Note 3).

5. Each tube must be prepared individually immediately prior to analysis. First, add 9 μL of the 5 μM hematin solution to each tube. This brings the concentration of heme to 22.5 nM, half the concentration of enzyme to be used (see Note 6).

6. Next, add pure COX enzyme to a final concentration of 45 nM. This concentration may be increased or decreased depending on the activity of the purified enzyme.

7. An enzyme-free control must be included to account for baseline oxidation of ABTS by free heme. This reactivity will be subtracted from each trace obtained.

8. Triturate the mixture and incubate at room temperature for no more than 1 min.

9. Transfer 1950 μL of the mix to a 3.5 mL absorption cuvette with a cell-stirring bar. The reaction should be stirred at ~250 RPM and maintained at 25°C.

10. Blank the spectrophotometer against the contents of the cuvette.

11. Add 50 μL of the desired peroxide solution to the cuvette and monitor the change in absorbance at 417 nm for

approximately 1 min. In order to obtain the intended kinetic data, a range of peroxide concentrations may be required. However, some peroxides may exhibit poor solubility at high concentrations (see Note 4).

12. The traces should be exported to a suitable graphing program such as MS Excel. The initial velocity is determined by a linear fit of the initial portion of the trace (Fig. 4a, b) during which time substrate consumption does not significantly affect reaction rate (see Note 7).

13. The velocity obtained ($\Delta[\text{Abs}]/\Delta t$) may be converted to a change in concentration of ABTS over time ($\Delta[\text{ABTS}]/\Delta t$), using the extinction coefficient of ABTS [ε_{mM} (417 nm) = 36] (6). Since the stoichiometry of the reaction of ABTS with peroxide

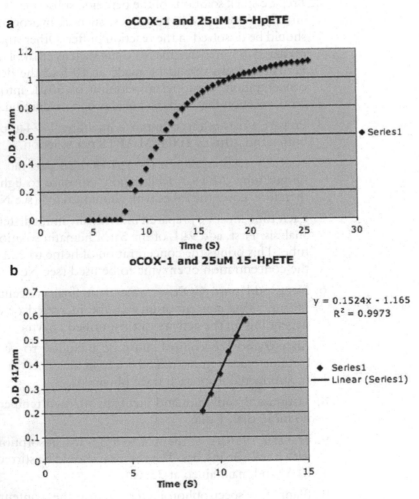

Fig. 4. A single determination of the turnover of 25 μM 15-HpETE via the peroxidase assay. Note the trace on the left displaying the entire reaction trajectory, with a loss of linearity after ~12 s. The spike in the trace at ~9 s is from the heat of dissolution from the addition of substrate to the reaction mixture (25 μM in 95% ETOH). Right, linear fit of the initial portion of the reaction with the slope representing the reaction rate.

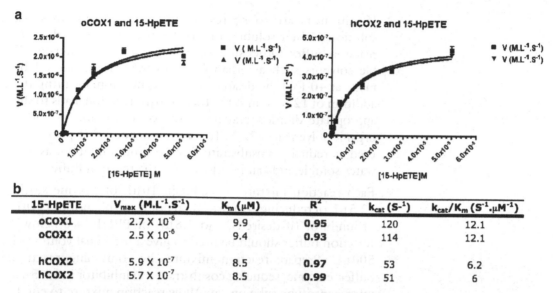

Fig. 5. (a) Determination of Michaelis–Menten curves for both oCOX-1 and hCOX-2 with 15-hydroperoxyeicosatetraenoic acid (15-HpETE) as a substrate. Each point on the velocity vs. [15-HpETE] was generated from a set of duplicates, with traces representative of those seen in Fig. 4. The nonlinear regression was generated using the graph pad program Prism (Version 4.0), fitting the points to the Michaelis-Menten equation ($V = V_{max}$ [ROOH]/K_m + [ROOH]) from which V_{max} and K_m were obtained. k_{cat} could be calculated ($k_{cat} = V_{max}/[E_a]t \rightarrow [E_a]t$ the total active enzyme). These values, along with the "goodness of fit" (R^2) are tabulated in (b).

is 2:1, the velocities obtained should be halved (9). It is also important to account for the nonenzymatic oxidation of ABTS by subtracting the rate obtained for the enzyme-free control from any experimental velocities (Fig. 5).

3.2. PPHP Assay

1. Prepare a 5 µM stock solution of COX enzyme in Tris buffer reconstituted with 2.5 µM hematin. Incubate on ice for at least 5 min before use. Adequate solution should be prepared to provide 10 µL per reaction mixture. The final enzyme concentration will be 100 nM. Depending on the specific activity of the enzyme, the concentration should be adjusted to give approximately 50% conversion of PPHP in the presence of 200 µM phenol, an established COX peroxidase reductant (see Note 8).

2. Prepare stock solutions of the reducing cosubstrate(s). Water-soluble compounds may be dissolved in the reaction buffer. Other stocks may be prepared in water miscible solvents such as ethanol or DMSO. Reducing cosubstrate stocks should be made at 40-fold the desired final concentration, ensuring that addition of 12.5 µL into the total 500 µL reaction leads to the appropriate dilution. In the absence of reducing cosubstrates to be tested, phenol may be used for all reaction mixtures. It may be added to the reaction buffer at a concentration of 200 µM.

3. If inhibitors are to be tested, prepare appropriate stock solutions. Water-soluble compounds may be dissolved in the reaction buffer. Other stocks may be prepared in water-miscible solvents such as ethanol or DMSO. Stocks should be made at 40-fold the desired final concentration, ensuring the addition of 12.5 µL into the total 500 µL reaction leads to the appropriate dilution and limits the concentration of added organic solvent to <2.5%. It is recommended that phenol be used as reducing cosubstrate when testing inhibitors, as it is water-soluble and can be added with the reaction buffer.

4. Each reaction mixture will contain 10 µL of enzyme stock, 12.5 µL of reducing cosubstrate (if not using phenol), 12.5 µL of inhibitor (if desired), and 1.8 µL of PPHP (see below). Reaction buffer should be used to give a total final volume of 500 µL. Prepare reaction mixtures by combining reaction buffer, enzyme, reducing cosubstrate and inhibitor in a 1.5 µL microcentrifuge tube on ice. Allow reaction mixture to equilibrate at 37°C for approximately 5 min.

5. A "no-enzyme" control should also be included to ensure the reduction of PPHP is enzyme-dependent and to account for any nonenzymatic reduction of PPHP (see Note 9).

6. To initiate the reaction, add PPHP to yield a final concentration of 100 µM. Since PPHP is supplied in a solution of 5 µg/µL (0.028 M), the addition of 1.8 µL results in the correct final concentration.

7. The reaction is terminated after 5 min by the addition of 500 µL of ice-cold quench solution followed by vigorous mixing on a Vortex mixer and storage on ice. The organic and aqueous layers are separated by centrifugation for 8 min at 4°C at 16,000 g using a bench-top centrifuge (see Notes 10 and 11).

8. The organic layer is removed, placed in a glass test tube, evaporated to dryness under nitrogen, and reconstituted in 300 µL of a 1:1 solution of methanol and water. The entire solution is then transferred to autosampler vials for HPLC analysis.

9. A PPHP standard should be prepared for HLPC analysis by the addition of 1.8 µL of PPHP to 300 µL of MeOH: H_2O (see Note 12).

10. Samples are separated by reverse phase liquid chromatography with UV detection at 254 nM. Samples are chromatographed utilizing a C8 reverse-phase column (described above) with a gradient elution beginning at 50% (v/v) methanol/water increasing to 90% methanol over 9 min and held at 90% methanol for 10 additional minutes at a flow rate of 0.20 mL/min (see Note 13). A sample chromatogram is shown in Fig. 6. A suggested injection volume is 10 µL.

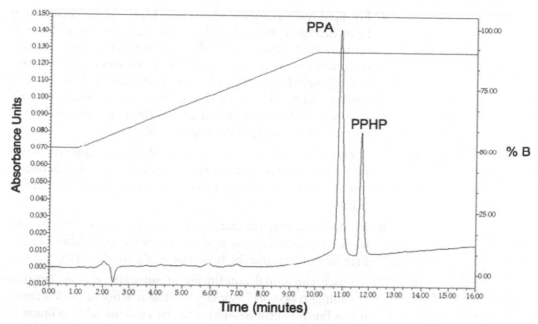

Fig. 6. High-performance liquid chromatogram for incubation of 100 nM mCOX-2 with 200 μM phenol and 100 μM PPHP using the method is described in the text. The percentage of solvent B (MeOH) is indicated by the grey line and the right Y-axis.

11. Peroxidase activity is measured by the percent conversion of PPHP to PPA using the equation: $[PPA]/([PPA]+[PPHP]) \times 100\%$. The concentration of the analytes is estimated from the area of the representative peak. When investigating the capability of a particular compound to act as reducing cosubstrate or inhibitor of peroxidase activity, activity can then be expressed as activity relative to control (no additional compound added) (see Note 14).

4. Notes

1. 100 mM sodium phosphate buffer at pH 7.8 can be used as a substitute.

2. This is a general assay that can be used to measure the peroxidase activity of many heme or nonheme peroxidases including: horseradish peroxidase, catalase, lactoperoxidase, cytochrome P450, and glutathione peroxidase (9).

3. ABTS should always be made fresh, kept on ice and out of direct light. Failure to do so results in autoxidation and decreased sensitivity of the assay. Freshly made solution should be pale yellow and will turn green if oxidation is taking place.

4. For lipid hydroperoxide substrates, high concentrations may form detergent micelles. This effectively decreases the concentration of substrate available for turnover, possibly leading to conditions where substrate is not in excess and/or introducing turbidity in the solution. This may lead to spurious kinetic results. The critical micelle concentrations of substrates should be established to determine the upper limit of concentrations of lipid hydroperoxide to use.

5. Peroxide substrates that are not soluble in aqueous media may be dissolved in any organic solvent that is miscible in water; however, organic solvent content should be kept at $\leq 2.5\%$.

6. Excess heme may generate exaggerated kinetics; a no enzyme control is critical for accurate determination of peroxidase kinetics. Free heme can cause the breakdown of peroxide (10).

7. To ascertain initial conditions of substrate excess, examine the initial trajectory (velocity) of the reaction and determine if it is linear over time. This may be established by a linear fit of initial reaction velocities in a graphing program. Generally, linear fits with "goodness of fit" (R^2) greater than 95% are best. Lower peroxide concentrations will yield fewer time points in the linear range as substrate consumption significantly reduces reaction velocity leading to a plateau in the absorbance trace.

8. The 2:1 ratio of peroxidase reductant to hydroperoxide substrate is chosen to reflect the stoichiometry of hydroperoxide reduction (9).

9. It is common to see a small amount (approximately 3–5%) of nonenzymatic conversion of PPHP to PPA.

10. As an alternative to quenching the reaction by addition of an organic solvent, the sample can be directly applied to reverse-phase solid-phase extraction (SPE) columns to extract PPHP and PPA as described by Weller et al. (9).

11. If desired, p-nitrobenzyl alcohol can be added to the organic phase after centrifugation and separation as a chromatographic standard and used for quantification (9).

12. To confirm the identity of the PPHP metabolite, PPA can also be purchased from Cayman Chemical and used as a chromatographic standard.

13. While a gradient chromatographic method is reported here, several laboratories have successfully performed isocratic HPLC analysis for the separation of PPHP and PPA with 56–65% (v/v) methanol/water as the mobile phase (9, 11). Additionally, the flow rate may be adjusted depending on the dimensions of the column.

14. The extraction efficiency can be determined by dividing the combined area of the peaks representative of PPHP and PPA in the sample reactions by the area of the peak in the PPHP standard. The extraction efficiency should be greater than 50%.

Acknowledgments

We would like to thank Carol Rouzer for a critical reading and editorial assistance.

References

1. Rouzer CA, Marnett LJ (2003) Mechanism of free radical oxygenation of polyunsaturated fatty acids by cyclooxygenases. Chem Rev 103:2239–2304

2. Hamberg M, Samuelsson B (1973) Detection and isolation of an endoperoxide intermediate in prostaglandin biosynthesis. Proc Natl Acad Sci USA 70:899–903

3. Van der Ouderaa FJ, Buytenhek M, Nugteren DH, Van Dorp DA (1977) Purification and characterisation of prostaglandin endoperoxide synthetase from sheep vesicular glands. Biochim Biophys Acta 487:315–331

4. Dietz R, Nastainczyk W, Ruf HH (1988) Higher oxidation states of prostaglandin H synthase. Rapid electronic spectroscopy detected two spectral intermediates during the peroxidase reaction with prostaglandin G2. Eur J Biochem 171:321–328

5. Karthein R, Dietz R, Nastainczyk W, Ruf HH (1988) Higher oxidation states of prostaglandin H synthase. EPR study of a transient tyrosyl radical in the enzyme during the peroxidase reaction. Eur J Biochem 171: 313–320

6. Childs RE, Bardsley WG (1975) The steady-state kinetics of peroxidase with 2, 2′-azino-di-(3-ethyl-benzthiazoline-6-sulphonic acid) as chromogen. Biochem J 145:93–103

7. Markey CM, Alward A, Weller PE, Marnett LJ (1987) Quantitative studies of hydroperoxide reduction by prostaglandin H synthase. Reducing substrate specificity and the relationship of peroxidase to cyclooxygenase activities. J Biol Chem 262: 6266–6279

8. Rowlinson SW, Crews BC, Lanzo CA, Marnett LJ (1999) The binding of arachidonic acid in the cyclooxygenase active site of mouse prostaglandin endoperoxide synthase-2 (COX-2). A putative L-shaped binding conformation utilizing the top channel region. J Biol Chem 274:23305–23310

9. Weller PE, Markey CM, Marnett LJ (1985) Enzymatic reduction of 5-phenyl-4-pentenyl-hydroperoxide: detection of peroxidases and identification of peroxidase reducing substrates. Arch Biochem Biophys 243: 633–643

10. Goldstein S, Meyerstein D, Czapski G (1993) The Fenton reagents. Free Radic Biol Med 15:435–445

11. Aronoff DM, Boutaud O, Marnett LJ, Oates JA (2003) Inhibition of prostaglandin H2 synthases by salicylate is dependent on the oxidative state of the enzymes. J Pharmacol Exp Ther 304:589–595

Study of Inhibitors of the PGH Synthases for which Potency Is Regulated by the Redox State of the Enzymes

Olivier Boutaud and John A. Oates

Abstract

Aspirin, salicylic acid, and acetaminophen are examples of drugs that inhibit the PGH synthases (cyclooxygenases) with potencies that vary remarkably between different cells. This results from the fact that the inhibitory action of these drugs is regulated by redox cycling of the enzymes, which is determined by the concentration of hydroperoxides in the enzyme proximity. The potency of these drugs is greatest in the presence of low concentrations of hydroperoxides. Accordingly, analysis of the action of these inhibitors requires consideration of assay conditions that take into account the factors that control hydroperoxide concentrations, both in studies with purified enzymes and in cells.

Key words: Cyclooxygenase, Assay, Hydroperoxide, Inhibitor, Acetaminophen, Aspirin, Salicylic acid

1. Introduction

The potency of acetaminophen, salicylic acid, and aspirin as inhibitors of the prostaglandin H synthases (PGHSs, cyclooxygenases) varies considerably between different cell types. This cellular selectivity is manifested in the human pharmacology of the drugs. For example, all three of these agents exert little or no antiinflammatory action at doses that are antipyretic and analgesic. Cellular selectivity is a consequence of the fact that inhibition of the PGHSs by these drugs is regulated by the oxidative state of the enzymes.

The bifunctional nature of the PGHSs provides a conceptual framework for consideration of the regulation of the action of these three drugs. The PGHSs contain both a heme-containing peroxidase and a cyclooxygenase catalytic site. Reduction of a hydroperoxide by the peroxidase oxidizes the heme to a protoporphyrin radical

Samir S. Ayoub et al. (eds.), *Cyclooxygenases: Methods and Protocols*, Methods in Molecular Biology, vol. 644,
DOI 10.1007/978-1-59745-364-6_7, © Springer Science+Business Media, LLC 2010

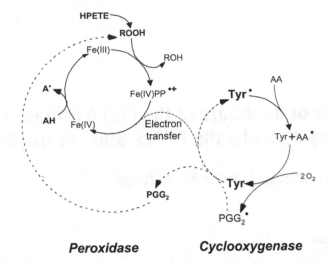

Peroxidase   **Cyclooxygenase**

Fig. 1. Catalytic mechanism of PGHSs. *Fe(IV)PP*•+ ferryl protoporphyrin IX radical cation, *Fe(III)* ferric enzyme, *Fe(IV)* ferryl protoporphyrin IX, *Tyr*• tyrosine 385 radical, *AH* reducing cosubstrate, *AA* arachidonic acid, *AA*• arachidonic acid radical, *PGG$_2$* prostaglandin G$_2$, *PGG$_2$*• prostaglandin G$_2$ with peroxyl radical on carbon 15, *ROOH* hydroperoxide substrate, *HPETE* hydroperoxyde of fatty acids.

cation that through intramolecular electron transfer generates a Tyr385 radical in the cyclooxygenase catalytic pocket (Fig. 1). Abstraction of hydrogen from arachidonic acid by the Tyr385 radical initiates the oxygenation that leads to the formation of prostaglandin G$_2$ (PGG$_2$). Thus, the PGHSs are catalytically active only after reduction of a hydroperoxide in the PGHS peroxidase site.

Among the hydroperoxide substrates of the PGHS peroxidase, some are rapidly reduced, thereby driving the heme to its higher oxidative state more efficiently than others. These substrates with a high turnover rate include fatty acid hydroperoxides, such as 12-hydroperoxyeicosatetraenoic acid (12-HPETE), peroxynitrite, and PGG$_2$, the product of the PGHS cyclooxygenase.

Acetaminophen: Acetaminophen inhibits both PGHS isoforms by reducing the protoporphyrin radical cation, thereby preventing the formation of the Tyr385 radical that is required for catalytic activity (1, 2). In effect, acetaminophen is competing with Tyr385 to provide the electron that reduces the protoporphyrin radical cation. This action of acetaminophen is antagonized by hydroperoxides that drive the formation of the protoporphyrin radical cation, so that acetaminophen is most effective at low concentrations of hydroperoxide and becomes less potent at higher concentrations of hydroperoxide. Thus, the efficacy of acetaminophen as a PGHS inhibitor is greatest in cells with a low concentration of hydroperoxide and is relatively less effective in cells, such as activated inflammatory cells, in which the concentration of hydroperoxide is high. This is the basis for the cellular selectivity of acetaminophen. A complementary mechanism by which acetaminophen acts is to lower the concentration of peroxynitrite by directly reducing it (3); by lowering the concentration of this PGHS peroxidase substrate, acetaminophen potency as an inhibitor of the PGHSs is increased.

The potency of acetaminophen as an inhibitor of the PGHSs is dependent on the concentration of both the enzyme and the substrate. Increased enzyme activity and/or increased substrate concentration will elevate the amount of PGG_2 formed, and this hydroperoxide product of the PGHS cyclooxygenase will antagonize the inhibitory effect of acetaminophen (Fig. 2). Thus, when comparing the potency of acetaminophen against the two PGHS isoforms, mutants of the PGHSs, or splice variants, it is essential in such in vitro experiments to utilize enzyme concentrations of equal catalytic activity and to employ the same concentration of arachidonic acid. Also, in studies on cells, the inhibitory effect of acetaminophen will be inversely related to the concentration of added arachidonic acid.

Earlier research on acetaminophen utilized the oxygraph to measure oxygen consumption as an indicator of catalytic activity. The very large concentrations of both the enzyme and the substrate required in experiments with the oxygraph produced high concentrations of PGG_2 that antagonized the action of acetaminophen and led to the erroneous conclusion that the IC_{50} was in the millimolar range, suggesting that acetaminophen was not an effective inhibitor of the PGHSs. The methods described here for

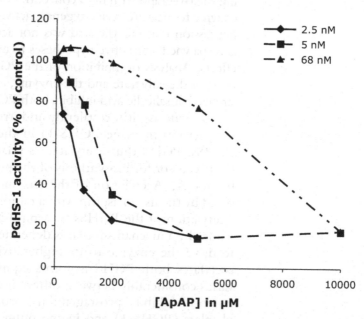

Fig. 2. Effects of PGHSs concentrations on the efficacy of acetaminophen. Inhibition of varying concentrations of PGHS-1 by acetaminophen was determined as described in Subheading 3.1 of this chapter. The concentrations of enzymes are indicated in the figure. The catalytic activity of PGHS-1 was 202 mol of arachidonic acid oxidized/min/ mol of enzyme. *ApAP* acetaminophen.

analysis of acetaminophen effect employ much lower concentrations of both the enzyme and the substrate.

Salicylic acid: Salicylic acid competitively inhibits the PGH synthases by blocking the access of arachidonic acid to the catalytic site in the enzymes. It does not act to reduce the protoporphyrin radical as is the case with acetaminophen. Its potency as a competitive inhibitor, however, also is regulated by the oxidative state of the enzymes. Thus, elevated levels of hydroperoxides will antagonize the inhibition of the PGHSs by salicylic acid, and salicylic acid is less potent as an inhibitor in cells that have high levels of hydroperoxide. The high levels of hydroperoxide formed during platelet activation, for example, render salicylic acid a poor inhibitor of thromboxane A_2 production in platelets that are activated. It is presumed that hydroperoxide-induced formation of the protoporphyrin radical prevents salicylate from binding in the catalytic channel of the PGHS cyclooxygenase, either by changing conformation of the channel or by altering the oxidative state of one of the residues with which salicylic acid binds.

Because of the effect of hydroperoxide concentration on salicylic acid potency, any comparison of the potency of salicylic acid between wild type and mutant enzymes, or between isoforms and splice variants, must be carried out with enzymes of equal catalytic activity. As was the case with acetaminophen, earlier work employing the oxygraph with high concentrations of arachidonic acid and enzyme to measure cyclooxygenase activity had led to an erroneous impression that salicylic acid was not acting as a PGHS inhibitor and spawned alternative hypotheses to explain its pharmacological effects. Analysis of inhibition of the PGHSs with low concentrations of the substrate and the enzyme, as described here, demonstrates that salicylic acid inhibits both PGHS isoforms.

Aspirin: Aspirin covalently and irreversibly acetylates a serine (Ser530 in ovine PGHS-1) in the catalytic pocket of the PGHSs. PGHS thus acetylated is unable to synthesize PGH_2, the precursor for biosynthesis of the prostaglandins and thromboxane A_2. Acetylation of the enzyme by aspirin can be measured by the use of aspirin with a radiolabel on the acetyl group. Acetylation of the PGHSs by aspirin is inhibited by hydroperoxides by a mechanism that is dependent on peroxidase activity to drive the enzyme to its higher oxidative state (4). Aspirin acetylates both PGHS isoforms equivalently, but its potency varies considerably between different cell types; for example, it potently inhibits prostaglandin/thromboxane production in platelets (PGHS-1) and in the pulmonary epithelial cell line, A549 (PGHS-2), but is far less potent in the inflammatory cell line, RAW264.7 (PGHS-2) which has high levels of lipid hydroperoxides.

2. Materials

2.1. Enzymatic Reaction Using Purified Enzymes

1. Ovine PGHS-1 or human recombinant PGHS-2 can be obtained from commercial sources such as Cayman Chemical (Ann Arbor, MI) or Oxford Biomedical Research, Inc. (Oxford, MI).

2. The working solution (500 μM) of hematin (H-3281, Sigma, MO) is prepared by dissolving 3.2 mg of hematin in 10 ml of dimethylsulfoxide (DMSO). After vortexing until complete dissolution of the hematin, the solution is kept at room temperature in an amber vial. In these conditions, the solution is stable for at least 1 year.

3. The working solution (1 μM) of [^{14}C] arachidonic acid (GE Healthcare Bio-Sciences Corp., NJ) is prepared right before being used for the reaction. 2.5 μl of [^{14}C] arachidonic acid (58.0 mCi/mmol) are diluted in 22.5 μl of 0.1 N NaOH. Then, 975 μl of Tris–HCl buffer pH 8.0 are added. The working solution is kept on ice until used.

4. Arachidonic acid (Nu-Chek Prep, Inc., MN) is dissolved in ethanol at a final concentration of 10 mg/ml and kept under argon at −20°C for several months.

5. A stock solution of butylated hydroxyanisol (BHA, Sigma, MO) 2 mg/ml, arachidonic acid 2 mg/ml is prepared by dissolving 4 mg of BHA in 1.6 ml of ethanol, to which 400 μl of 10 mg/ml arachidonic acid is added. The solution is kept under an argon atmosphere at 4°C in an amber vial until used. In these conditions, the solution is stable for several months.

6. Preparation of the terminating solution: The solvent for the terminating solution is prepared in advance by mixing 450 ml of diethyl ether, 60 ml of methanol and 15 ml of citric acid 4 M in water and stored at 4°C in a glass bottle. Right before the experiment, the required volume of terminating solution is reconstituted by adding 100 μl of 2 mg/ml BHA/arachidonic acid for every 5 ml of terminating solution solvent.

7. Stock solution of acetaminophen (Sigma, MO) is prepared by dissolving 15 mg in 1,000 μl of ethanol (100 mM). The working solutions are prepared by serial dilution of the stock solution with ethanol to obtain solutions that are 100 times more concentrated than the concentration desired in the final reaction.

8. Stock solution of salicylic acid (Sigma, MO) is prepared by dissolving 16 mg in 1,000 μl of ethanol (100 mM). The working solutions are prepared by serial dilution of the stock solution with ethanol to obtain solutions that are 100 times more concentrated than the concentration desired in the final reaction. The solutions are discarded after the experiment is done.

9. Stock solution of aspirin (Sigma, MO) is prepared by dissolving 18 mg in 1,000 μl of ethanol (100 mM). The working solutions are prepared by serial dilution of the stock solution with ethanol to obtain solutions that are 100 times more concentrated than the concentration desired in the final reaction. The solutions are discarded after the experiment is done.

10. [acetyl-1-^{14}C] Aspirin (American Radioactive Chemicals, MO; 55 Ci/mol, 1.82 mM) is kept in toluene at 4°C. 7.5 μl of [^{14}C] aspirin represent 13.6 nmol, 0.5 μCi.

11. Platelet wash buffer (potassium phosphate buffer, 113 mM NaCl, 5.5 mM glucose, pH 6.5): Weigh 6.6 g of NaCl, 0.75 g of K_2HPO_4 anhydrous (Sigma, MS), 0.61 g Na_2HPO_4 anhydrous (Sigma, MS), 3.37 g of NaH_2PO_4 monohydrate (Sigma, MS), and 0.99 g of glucose (Sigma, MS). Dissolve in 800 ml of water and adjust pH to 6.5 with 0.2 N NaOH. Adjust volume to 1 L with water.

12. Platelet suspension buffer (potassium phosphate buffer, 109 mM NaCl, 5.5 mM glucose, pH 7.5): Weigh 6.37 g of NaCl, 0.75 g of K_2HPO_4 anhydrous (Sigma, MS), 1.15 g of NaH_2PO_4 monohydrate (Sigma, MS), and 0.99 g of glucose (Sigma, MS). Dissolve in 800 ml of water and adjust pH to 7.5 with 0.2 N NaOH. Adjust volume to 1 L with water.

13. Four series of 1.5 ml microcentrifuge tubes (Cat. # 05-408-129, Fisher Scientific) are prepared before starting the reaction: two series are used for carrying out the oxidation reactions (enzymatic and control), the other are used to collect the organic phase containing the extracted products of the reactions.

14. Thin layer chromatography (TLC) plates Partisil® LK6D are from Whatman International, Ltd., England. The filter paper is from BioRad.

15. The chromatographic solvents are prepared as follows and are stored at 4°C in a tight glass bottle until ready to use. Ethyl acetate (360 ml), isooctane (200 ml), water (400 ml), and acetic acid (8 ml) are mixed in a separatory funnel and shaken vigorously. The separatory funnel is inverted and the valve is opened to let the gas escape. The process is repeated once. The separatory funnel is then placed on a stand and the two phases allowed to separate. The top stopper is removed and the aqueous phase (bottom layer) is discarded according to the chemical waste regulations. Care is taken not to keep any water in the organic phase by discarding a couple of ml of it. The organic phase is collected and is used for the chromatography.

16. A TLC radioactivity scanner is used to monitor the oxidation of arachidonic acid (AR-2000, Bioscan, Inc., DC). Radioactive regions of interest are integrated using the software Win-Scan (Bioscan, Inc., DC).

2.2. Preparation of Human Washed Platelets

1. The disposable 5 ml polypropylene columns (29922) are from Thermo Scientific, IL.

2. Sepharose 2B is from GE Healthcare Bio-Sciences (17-0130-01).

3. Sodium azide and citric acid are from Sigma, MS.

4. Butterfly needles gauge $19^3/_4$ (B387359) and sodium citrate (BP327) are from Fisher Scientific.

5. Red cap vacuum tubes (367814) are from BD Vacutainer.

6. Sterile plastic syringes STRL Luerlock (BD309653) are from VWR.

2.3. Inhibition of Cyclooxygenases in Cells

1. The human colon cancer cell line HCA-7, which was established from a patient with well-differentiated mucoid adeno-carcinoma of the colon (5), was kindly provided by Dr. Susan Kirkland (University of London). A-549 (CCL-185) and RAW264.7 (TIB-71) are from ATCC.

2. DMEM (12800-017), fetal bovine serum (16140-071) and antibiotics (15240-062) are from Invitrogen.

3. Interleukin 1b (IL-1β, 19401), endotoxin (LPS, L8274), and interferon-g (INF-γ, I4777) are from Sigma, MS.

4. The nylon filters 0.22 μm (28199-303) are from VWR.

5. The 6-well plates (087721B) are from Fisher Scientific.

2.4. Analysis of Prostanoids by Gas Chromatography Mass Spectrometry

1. N,N-diisopropylethylamine (DIPEA; D125806), 2,3,4,5,6-pentafluorobenzyl bromide (PFB-Br; 101052), methoxylamine hydrochloride (226904), pyridine anhydrous (270970), and phosphomolybdic acid spray reagent (P4869) are from Sigma, MS. Trimethylchlorosilane and N,O-bis(trimethylsilyl)trifluoroacetamide (TMCS-BSTFA; 33154-U) is from Supelco (Bellefonte, PA).

2. Prostaglandin D_2 methyl ester (10008385), prostaglandin E_2 methyl ester (14011), and prostaglandin $F_{2\alpha}$ methyl ester (16011) are from Cayman.

3. The transfer pipettes (14670-455) and the vacuum box (SPE-12G column processor, JT7018-0) are from VWR.

4. The glass reactivial (22021822) and the undecane (14066-1000) are from Fisher Scientific.

5. The C18 solid phase extraction cartridges (WAT036575) are from Waters.

6. The gas chromatography tandem mass spectrometer (5975 Inert MSD/DS) and the silica column DB-1701 (122-3762) are from Agilent Technologies, DE.

3. Methods

Prior to the evaluation of the effect of inhibitors on purified or expressed PGHS, the optimal concentration of enzyme to be employed in the analysis must be determined. The catalytic activity of purified or expressed PGHSs varies from one preparation to the other, and therefore cannot be predicted from the number of moles of protein. The concentration of the enzyme employed should be sufficient to produce an amount of oxygenation of arachidonic acid that can be measured accurately, and at the same time to yield a reaction rate that approaches linearity. This is achieved by measuring the extent of oxygenation produced by a range of enzyme concentrations in the assay described in Subheadings 3.1 and 3.2 below, and from these data determining an enzyme concentration in the reaction mixture that yields between 25% and 40% oxygenation of the AA substrate; this may be designated as the *optimal enzyme concentration* (see Note 1).

3.1. Determination of the Amount of PGHS to Be Employed in Analysis of the Potency of Inhibitors

1. To establish the *optimal enzyme concentration*, it is necessary to evaluate a range of enzyme concentration to determine a concentration that yields 25–40% oxygenation of the added arachidonic acid. This is accomplished by testing this range of PGHS concentrations in the enzymatic assay as described in Subheading 3.3, with the exception that steps 2–6 of Subheading 3.3 are substituted with the steps described below.

2. Six PGHS concentrations covering the range of 2–16 nM final concentration can be evaluated. This range will include the *optimal enzyme concentration* for most enzyme preparations; if not, then concentrations outside this range must be tested.

   For the enzyme reaction, tubes are set up on the basis of analysis of each PGHS concentration in duplicate in comparison with duplicate controls in which the enzyme is omitted. Thus, there will be four tubes for each enzyme concentration. Set up 24 tubes for six concentrations and also a second series of 24 tubes labeled identically to collect the products of the extraction from each reaction tube.

3. No inhibitor is employed in the studies to determine the *optimal enzyme concentration*.

4. Reconstitute the PGHS in buffer (Tris–HCl 100 mM, pH 8.0, 500 µM phenol) to yield a stock solution with a final enzyme concentration of 1.0 µM. Add two molar equivalents of hematin. Mix and place on ice for at least 30 min before using in the reaction.

5. Prepare the working solutions of the PGHS in six concentrations that will yield final concentrations in the reaction

mixture of 2, 4, 6, 8, 12, and 16 nM. This is accomplished by adding 2, 4, 6, 8, 12, and 16 μl of the PGHS stock solution to their respective tubes (1 μl stock solution for each nM final concentration) and bringing the final volume in all of the tubes to 900 μl with buffer (Tris–HCl 100 mM, pH 8.0, 500 μM phenol). This constitutes the series of six working solutions. Incubate the tubes at 37°C for 2 min.

6. Prepare the working solutions for the control incubations (no enzyme, see Note 2). In six Eppendorf tubes, place 900 μl of buffer (Tris–HCl 100 mM, pH 8.0, 500 μM phenol).

3.2. Determination of the Specific Activity of the PGHSs

From an experiment that employs the *optimal enzyme concentration*, the *specific activity* of the PGHS preparation, expressed as pmol arachidonic acid oxygenated/min/pmol PGHS, can be determined. In each reaction tube, there is 100 pmol of arachidonic acid. Therefore, the percent of oxygenated arachidonic acid is equivalent to the pmol of arachidonic acid oxygenated in 8 s. Calculation of the nmol of PGHS is based on 0.2 ml of the reaction mixture.

Knowing the specific activity, the number of *catalytic units* can be calculated (1 catalytic unit = pmol of enzyme that catalyzes oxygenation of 1 pmol of arachidonic acid/min). To compare the effect of inhibitors on the different PGHS isoforms, or on mutants or splice variants, it is necessary to make the comparison on the amounts of enzyme that have approximately the same number of *catalytic units* (see Note 1).

This is illustrated in Fig. 2, which shows that the efficacy of acetaminophen decreases as the enzyme concentrations increase from 5 to 65 nM.

3.3. Measurement of the Effects of Acetaminophen or Salicylic Acid on the Oxidation of Arachidonic Acid by Purified PGHSs

The following protocol is written for an experiment where five concentrations of inhibitor and a vehicle control (no inhibitor) are tested in the presence of the optimal enzyme concentration for the PGHS preparation (see Note 1).

1. Place the desired number of TLC plates in a tank of acetone about 60 min before starting the experiment (see Note 3).

2. Each reaction condition is carried out in duplicate and in two parallel series in which each enzymatic reaction is run together with a control in which the enzyme is omitted (see Note 2). Thus, for each inhibitor concentration, four tubes need to be labeled: two oxidations and two controls. When this is done, another series of tubes labeled identically should be prepared to collect the products of extraction of each condition. So, for five concentrations of inhibitor and a vehicle control (six reaction conditions) there should be a total of 48 tubes.

3. Prepare the working solutions of inhibitor by serial dilution of the 10 mM stock with ethanol. Each working solution

should be 100-fold more concentrated than the final drug concentration expected in the enzymatic reaction.

4. Reconstitute the enzymes by mixing the volume of stock solution necessary to have a concentration of enzyme of 1 μM in Tris–HCl 100 mM, pH 8.0, 500 μM phenol (see Note 1). Add two molar equivalents of hematin. Mix and incubate on ice for 30 min.

5. Prepare a working solution of the enzyme by mixing the number of μl of reconstituted enzyme that corresponds to the number of nM of the *optimal enzyme concentration* (e.g. for an optimal enzyme concentration of 5 nM, add 5 μl of reconstituted enzyme), 10 μl of inhibitor stock solution (see Note 4) and a volume of Tris–HCl 100 mM, pH 8.0, 500 μM phenol to bring the total volume of the working solution to 900 μl. Incubate the tubes at 37°C for 2 min.

6. Prepare the working solutions of control (no enzyme, see Note 2). In six Eppendorf tubes, mix 445 μl of Tris–HCl 100 mM, pH 8.0, 500 μM phenol, and 5 μl of inhibitor working solution in ethanol. Incubate the tubes at 37°C for 2 min.

7. Prepare the working solution of [^{14}C] arachidonic acid as described in the Subheading 2.1.3. Keep on ice until used to prevent oxidation.

8. Prepare the terminating solution by mixing 500 μl of 2 mg/ml BHA/arachidonic acid with 25 ml of terminating solution solvent as described in subheading 2.1.6. Keep on ice.

9. Put the empty reaction tubes in the water bath at 37°C. *Each reaction is done individually.* Add 20 μl of [^{14}C] arachidonic acid in the tube and immediately start the reaction by adding 180 μl of the working solution of enzyme that already has been incubating to achieve a temperature of 37°C. After 8 s, stop the reaction by adding 400 μl of terminating solution (see Note 5). Close the tube, vortex briefly and start the next reaction.

10. When the reactions have been completed, vortex all the tubes again and transfer 200 μl of the organic phase (top layer) to the corresponding empty tube. Make sure to avoid pipetting any water, as this would interfere with the chromatography (see Note 6).

11. Take out the TLC plates from the acetone tank and let them dry under the hood (see Note 3).

12. Load the organic extract on the plates making sure that you do not load any material at less than 0.5 cm from the bottom of the plate (see Note 7). Make sure the material is completely dry before starting the chromatography.

13. A glass tank is prepared by putting a sheet of filter paper (see Note 8) and the volume of chromatographic solvent that

gives a depth of solvent of 2 mm in the tank (see Note 9). The plates are put inside, the tank is closed with its lid and the plates are left in the tank until the solvent front migrates to the top of the plates. This step is performed at 4°C (in a cold room) and lasts around 1 h. The plates are then removed from the tank and dried in a chemical hood before scanning the radioactivity with the Bioscan AR-2000.

14. Regions of interest are integrated using Win-Scan following the instructions of the manufacturer. For each enzymatic condition, the percent of oxidized fatty acids is determined in the control reaction and subtracted from the percent of oxidized fatty acid determined in the oxidation reactions. The difference represents the percent of oxidized arachidonic acid catalyzed by PGHS (see Note 10).

3.4. Measurement of the Effects of Aspirin on the Oxidation of Arachidonic Acid by Purified PGHSs

Aspirin inhibits PGHS enzymes by irreversibly acetylating a serine residue in the cycloxygenase site. In contrast to other inhibitors of PGHS, this occurs before the enzyme is catalytically active. Acetylation is inhibited by the hydroperoxide, PGG_2, which is the product of arachidonic acid oxygenation in the cyclooxygenase site of the enzyme (4) and perhaps also by competition with arachidonic acid in the active site. For this reason, inhibition of PGHS by aspirin requires preincubation of the reconstituted enzyme with aspirin for 40 min before addition of the substrate.

In this section of the chapter, we describe the modifications of the previously described protocol that are required when studying PGHS inhibition by aspirin.

The first four steps are identical with the protocol for acetaminophen or salicylic acid. Steps 5 and 6 are modified as follows to allow for the acetylation to occur prior to the exposure to arachidonic acid.

Prepare a working solution of the enzyme by mixing the number of μl of reconstituted enzyme that corresponds to the number of nM of the *optimal enzyme concentration* (e.g. for an optimal enzyme concentration of 5 nM, add 5 μl of reconstituted enzyme), 10 μl of aspirin working solution in ethanol (see Note 4) and a volume of Tris–HCl 100 mM, pH 8.0, 500 μM phenol to bring the total volume of the working solution to 900 μl. Incubate the tubes at 37°C for 40 min.

Prepare the working solution of control (no enzyme, see Note 2). In six Eppendorf tubes, mix 445 μl of Tris–HCl 100 mM, pH 8.0, 500 μM phenol, and 5 μl of aspirin working solution in ethanol. Incubate the tubes at 37°C for 40 min.

Steps 7–14 are identical to the ones for acetaminophen and salicylic acid. The exact same precautions are recommended for the entire protocol.

3.5. Acetylation of the Purified PGHSs by Aspirin

To study PGHS acetylation by aspirin, the time-course of acetylation is determined over a range of 10 min and the rate of acetylation is calculated based on the amount of radioactive acetate incorporated in the protein.

The following protocol is an example that was written for an experiment in which final concentrations of $1.5\,\mu M$ of PGHS-1 and $2.7\,\mu M$ of PGHS-2 were used in order to provide equivalent catalytic activity. The amount of enzyme(s) to be employed with other batches of enzymes will need to be calculated accordingly to subheading 3.1 (see Note 11). The amount of enzyme used for this protocol is higher than that used for the activity experiments as measuring the radioactive acetate covalently bound to the enzyme by autoradiography requires more protein.

1. Each reaction condition is carried out in duplicate. Enough reaction mixture is prepared for six time points: 0, 2, 4, 6, 8, and 10 min. Prepare and label tubes with $5\,\mu l$ of SDS-PAGE loading buffer. Therefore, for each concentration of aspirin studied there should be a total of 13 tubes: 1 tube for incubation of the enzyme with aspirin at 37°C and 12 tubes containing $5\,\mu l$ of SDS-PAGE loading buffer for the different time points in duplicates. Keep the 12 tubes on the bench next to the water bath used for the experiment.

2. Prepare a water bath or a heating block at 70°C for terminating the reaction.

3. Reconstitute the enzyme by mixing the volume of stock solution necessary to have a final concentration of $1.5\,\mu M$ of PGHS-1 or $2.7\,\mu M$ of PGHS-2 with Tris–HCl 100 mM, pH 8.0, $500\,\mu M$ phenol (see Note 11). Add two molar equivalents of hematin. Mix and incubate this solution of enzyme on ice for 30 min.

4. Incubate the enzyme solution at 37°C for 2 min.

5. Mix in an Eppendorf tube $5\,\mu l$ of Tris–HCl 100 mM, pH 8.0, $500\,\mu M$ phenol, and $7.5\,\mu l$ of [acetyl-1-^{14}C] aspirin. Put in the water bath at 37°C and immediately add $137.5\,\mu l$ of the enzyme solution incubated at 37°C.

6. At each time point, take out $20\,\mu l$ of the reaction and mix with $5\,\mu l$ of SDS-PAGE loading buffer. Put the tube at 70°C for 10 min (see Note 12).

7. At the end of the experiment, the samples can be stored at –20°C until ready to be analyzed. The samples are loaded in a 10% polyacrylamide gel and after electrophoresis, the proteins are stained with Coomassie blue (250 mg/l in water/methanol/acetic acid at 5:4:1 v/v). After destaining with the same solvent system, the gels are dried in a gel dryer and the radioactivity associated with PGHS is determined by autoradiography using PhosphorScreen (Amersham Biosciences, Piscataway, NJ) for 5 days. The data are analyzed on Typhoon

9400 using the software Typhoon Scanner Control (Amersham Biosciences, Piscataway, NJ) following the manufacturer's instructions. The radioactivity associated with each band is quantified using the software ImageQuant 5.2 (Amersham Biosciences, Piscataway, NJ) and is expressed as a percentage of the intensity of the band at 10 min.

3.6. Inhibition of PGHS-1 in Human Washed Platelets

1. Preparation of the sepharose 2B is done at least 24 h before being used for the chromatography. 50 ml of Sepharose 2B slurry is poured in a 50 ml screwing cap plastic tube (see Note 13). The sepharose is allowed to pack by gravity and the supernatant (50% ethanol in water) is discarded. The beads are resuspended in one volume of wash buffer (see Subheading 2) and allowed to pack again by gravity. The supernatant is discarded and the beads are resuspended in 0.5 volume of wash buffer containing 0.02 % of sodium azide (see Note 14). The beads are kept at 4°C until poured in the columns.

2. Columns of Sepharose 2B are prepared in the morning of the blood draw. Washed sepharose 2B beads are resuspended by gentle inversion of the tube and degassed under vacuum for 1 min. The slurry is poured into a polypropylene column up to the top (8.5 ml). The beads are allowed to pack by gravity and then the cap is removed from the bottom opening and the buffer is allowed to flow (see Note 15). The column is washed with 10 ml of wash buffer taking care not to let the column dry. The cap is put back at the bottom opening until the platelets are ready to be washed.

3. Written informed consent is obtained from all study participants. Blood is collected into vacuum tubes from a peripheral vein with a 19 ¾-gauge needle. The first 3 ml of blood is discarded (see Note 16). The volume of blood needed for the different analyses is collected in plastic syringes (see Note 17) containing 0.1 volume of 3.8% sodium citrate. Citrated whole blood specimens should be kept at room temperature with minimal agitation until used (see Note 18).

4. Preparation of washed human platelets: 25 ml of whole blood collected as described above is transferred to 15 ml Falcon tubes, which are then centrifuged at $190 \times g$ for 20 min at room temperature. The supernatant, platelet-rich plasma (PRP), is transferred to a fresh plastic tube (see Note 19) and acidified to pH 6.4 with citric acid 0.15 M (see Note 20). The acidified PRP is kept at room temperature for 10 min and then centrifuged at $800 \times g$ for 10 min at room temperature. The supernatant is discarded and the pellet (platelet) is resuspended in 400 µl of wash buffer. The resuspended platelets are washed by gel filtration chromatography at room temperature using Sepharose 2B equilibrated with the same buffer (see Note 21). The eluate is discarded until the platelets

start to be eluted (the eluate becomes milky as the cells are coming off the column). At this time, the cells are collected in a fresh tube until the eluate starts to clear (see Note 22). The eluted platelets are counted using a Z1 Dual Threshold particle counter (Beckman-Coulter) with a 70 µl probe and then diluted with suspension to achieve a final concentration of 300,000 platelets per µl.

5. The following protocol is written for six experimental conditions: one control (no inhibitor) and five concentrations of aspirin. Prepare two series of tubes. One series of six tubes for the reactions and one series of 12 tubes for the extraction.

6. In the extraction tubes, add 200 µl of ethyl acetate and 30 µl of 1 N HCl. Keep on ice.

7. Prepare the working solutions of aspirin by serial dilution of the 10 mM stock in ethanol. Each working solution should be ten times more concentrated than the final drug concentration expected in the enzymatic reaction.

8. Prepare the working solution of arachidonic acid by diluting the 1 mg/ml stock solution in ethanol. Mix 3 µl of stock solution with 17 µl of 0.2 N NaOH. Add 180 µl of Tris–HCl buffer 100 mM pH 8.0 for a final concentration of 50 µM. Keep on ice until used to prevent oxidation.

9. Put your reaction tubes in the water bath at 37°C. Add 91 µl of washed platelets and 5 µl of the working solutions of aspirin. Incubate at 37°C for 40 min.

10. Start the reaction by adding 4 µl of arachidonic acid 50 µM (final concentration of 2 µM, see Note 23). Incubate at 37°C for 15 min.

11. At the end of the reaction, take 20 µl of reaction mixture and add to the corresponding extraction tube containing ethyl acetate and HCl. Close the tube and vortex quickly.

12. When the 12 reactions have been completed, add 10 µl of [^{2}H$_4$] thromboxane B$_2$ as internal standard (2 ng). Vortex all the tubes again and centrifuge at 2,000 × g for 2 min. Transfer the organic phase (top layer) to an empty glass culture tube. Make sure to avoid pipetting any water. Add 1 ml of ethyl acetate to each reaction tube and extract the prostanoids again. Combine the organic phase with the previous extraction (see Note 24).

3.7. Analysis of Thromboxane B$_2$ by Gas Chromatography Mass Spectrometry

1. Dissolve 120 mg of methoxylamine hydrochloride in 4 ml of water.

2. Evaporate the ethyl acetate extract to dryness (from step 12 in Sub-heading 3.3) under a stream of nitrogen. Immediately resuspend dry residue in 75 µl of acetonitrile. Vortex.

3. Add 300 µl of methoxylamine hydrochloride in water. Incubate at room temperature for 30 min.

4. Extract the derivatized prostaglandins by adding 1 ml of dichloromethane. Pull up the entire reaction mixture in a narrow tip plastic transfer pipette three times. Pull up entire reaction mixture in the transfer pipette again and let layers separate in the pipette.

5. Transfer the lower layer (dichloromethane) into a clean reactivial (see Note 25) and empty the top layer (aqueous) back into the extraction tube.

6. Repeat steps 4 and 5. Combine the organic phases (see Note 24). Discard the aqueous phase.

7. Evaporate the dichloromethane to dryness under a nitrogen stream (see Note 26). Resuspend the dry residue in 15 µl of N,N-diisopropylethylamine (DIPEA). Add 30 µl of 12.5% 2,3,4,5,6-pentafluorobenzyl bromide (PFB-Br) in anhydrous acetonitrile. Vortex. Keep at room temperature for 30 min (see Note 27).

8. Evaporate the solvents to dryness under a nitrogen stream and resuspend dry residue in 50 µl of methanol/chloroform (1:1 v/v). Vortex (see Note 24).

9. Take out the TLC plates from the acetone tank and let them dry under the hood (see Note 3).

10. Load the organic extract on the plates making sure that you do not load any material at less than 0.5 cm from the bottom of the plate (see Note 7). Load on a separate plate 5 µg of each of the methyl esters (Me-) of PGE_2, PGD_2, and $PGF_2\alpha$ as standards. Make sure the material is completely dry before starting the chromatography.

11. A glass tank is prepared by putting a sheet of filter paper (see Note 8) and between 50 and 100 ml of ethyl acetate/methanol (49:1 v/v) in it (see Note 9). The plates are put inside with the spots at which the samples are loaded at the bottom end of the plates. The tank is closed with its lid and the plates are left in the tank until the solvent front migrates to the top of the plates. This step lasts around 1 h. The plates are then removed from the tank and dried in a chemical hood.

12. The plate containing the standards is sprayed with phosphomolybdic acid under a hood. The methyl ester prostaglandins are revealed by heating the plate with a heat gun. Care should be taken not to overheat the plates as the silica would burn and the plates would break.

13. The regions of interest are determined by the position of the standards on the plate. Place a plate in a fume hood on a weighing paper that is creased by having been folded in two

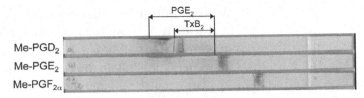

Fig. 3. TLC plate of methylester standards developed with phosphomolybdic acid. The positions used to determine the regions to scrape are indicated with arrows.

(see Note 28). Each sample lane is scraped between 0.5 cm above Me-PGE$_2$ and the leading edge of Me-PGD$_2$ (see Fig. 3). The silica collected on the weighing paper is then transferred into an Eppendorf tube.

14. Add 1 ml of ethyl acetate, vortex and centrifuge at 2,000×g for 6 min. Transfer supernatant in a new Eppendorf tube (see Note 24) making sure *NOT* to transfer any silica particles.

15. Evaporate ethyl acetate under nitrogen stream to *COMPLETE* dryness (see Note 29). Add 20 µl of dry pyridine and 30 µl of trimethylchlorosilane and N,O-*bis*(trimethylsilyl) trifluoroacetamide (TMCS-BSTFA). Vortex. Incubate at 37°C for 1 h or at room temperature overnight (see Note 30).

16. The solvents are evaporated to dryness under a nitrogen stream, the residue is solubilized in 20 µl of undecane, immediately transferred to a GC/MS vial that contains a glass insert to permit small volume sampling (see Note 31) and analyzed by GC/MS (see Note 32).

3.8. Inhibition of PGHSs in Cells in Culture

1. Human lung cancer A549 cells are grown to confluence at 37°C, 5% CO$_2$ in DMEM, 10% FBS, penicillin 100 U/ml, streptomycin 100 U/ml, and amphotericin B 250 ng/ml in 6-well plates. At this time, cells are activated with IL-1β at 1 ng/ml for 24 h. The medium is changed to 2 ml of serum-free DMEM before the inhibition experiment.

2. Colorectal adenocarcinoma cells, HCA-7, are grown to confluence at 37°C, 5% CO$_2$ in DMEM, 10% FBS, penicillin 100 U/ml, streptomycin 100 U/ml, and amphotericin B 250 ng/ml in 6-well plates. Because HCA-7 overexpresses PGHS-2 constitutively, they do not need to be activated. The medium is changed to 2 ml of serum-free DMEM before the inhibition experiment.

3. Mouse monomacrocytic RAW 264.7 cells are grown until they are 80% confluent at 37°C, 5% CO$_2$ in DMEM, 10% FBS, penicillin 100 U/ml, streptomycin 100 U/ml, and amphotericin B 250 ng/ml in 6-well plates. After changing the medium with new serum-free DMEM, the cells are activated with LPS 10 µg/ml and IFNγ 10 U/ml for 16 h.

The medium is changed to 2 ml of fresh serum-free DMEM before the inhibition experiment.

4. The different compounds used to activate the cells are prepared freshly by adding the volume of stock solution needed to the volume of cell culture medium necessary for the experiment such as to get the final concentrations indicated above for each cell type. The medium is then filtered sterile on a 0.22 μm nylon filter before adding it to the cells.

5. Prepare plastic tubes with caps that can contain volumes up to 4 ml.

6. Prepare the stock solutions of inhibitors by dissolving 18.0 mg of aspirin in 1 ml of ethanol (100 mM stock), 14 mg of salicylic acid in 1 ml of ethanol (100 mM stock) or 151 mg/ml of acetaminophen in 1 ml of ethanol (1 M stock, see Note 33). Prepare the working solutions of inhibitors by serial dilution of the stock solution in ethanol. Each working solution should be 160 times more concentrated than the final drug concentration expected in the cell culture well.

7. Prepare the working solution of $[^2H_8]$ arachidonic acid by mixing 20 μl of the stock solution (1 mg/ml) with 180 μl of ethanol for a concentration of 330 μM (see Note 34). Keep on ice until used to prevent oxidation.

8. Add 4.0 μl of working solution of the inhibitor to the freshly changed serum free DMEM. Put the plates back in the incubator for 20 min (acetaminophen or salicylic acid) or 40 min (aspirin).

9. Add 12.5 μl of $[^2H_8]$ arachidonic acid 330 μM to each well (final concentration of 2 μM, see Note 23). Put the plates back in the incubator for 15 min.

10. Collect the medium and transfer it to the plastic tubes.

11. Add 10 μl of $[^2H_4]$ prostaglandin E_2 as internal standard (2 ng). Freeze at −20°C until ready for analysis by mass spectrometry.

3.9. Analysis of Prostaglandin E₂ by Gas Chromatography Mass Spectrometry

1. Dissolve 120 mg of methoxylamine hydrochloride in 4 ml of water.

2. C18 Sep-Pak cartridges are conditioned by washing them with 10 ml of methanol followed by 10 ml of pH 3 water. The cartridges are placed on a vacuum box and the vacuum is set to −10 kPa.

3. Cell media are thawed and the pH is adjusted to 3 with 1 N HCl (see Note 35).

4. The samples are loaded on the Sep-Pak cartridge. The column is washed with 10 ml of water, followed by 10 ml of heptane (see Note 36). The prostanoids are eluted in 10 ml of ethyl acetate/heptane (1:1 v/v).

5. Evaporate the solvents under nitrogen stream to dryness. Resuspend dry residue in 75 µl of acetonitrile. Vortex.

6. The prostanoids are derivatized following steps 2–12 of the protocol described for thromboxane in Subheading 3.5 above.

7. The regions of interest are determined by the position of the standards on the plate. Place a plate in a fume hood on a weighing paper that is creased by having been folded in two (see Note 28). Each sample lane is scraped between 0.5 cm above PGE_2-Me and 2 cm above PGD_2-Me (see Fig. 3). The silica is collected and transferred into an Eppendorf tube.

8. The rest of the protocol is similar to steps 14–16 of the protocol described for thromboxane B_2 in Subheading 3.5 above. The exact same precautions are recommended for the entire protocol.

4. Notes

1. To accurately compare the potency of inhibitors between isoforms, splice variants or mutants, the amounts of the enzymes employed in the reaction must have approximately equal catalytic activity. First, the specific activities of the two enzyme preparations used must be assessed, and then the concentration of the enzymes must be adjusted accordingly. As an example, in our investigations of acetaminophen (1) we compared PGHS isoforms that had specific activities of 202 mol arachidonic acid oxidized/min/mol of PGHS-1 and 109 mol arachidonic acid oxidized/min/mol of PGHS-2. The specific activities of PGHS-1 and -2 were determined and the amount of enzymes employed in the comparison was adjusted to have approximately equal number of catalytic units. The amount of enzyme that yields between 25 and 40% oxygenation of 0.5 µM arachidonic acid during 8 s of reaction time is designated as the *optimal enzyme concentration*.

2. Because arachidonic acid is prone to oxidation in contact with air, a parallel experiment is done in which everything is mixed together with the exception of the enzyme. The samples are analyzed the same way as the enzymatic reactions and the percent of oxidation in these control reactions are determined and subtracted from each equivalent enzymatic reaction. This corrects for any nonenzymatic oxidation caused by the presence of air and other chemicals present in the enzymatic reaction mixture and allows expressing the results as percent of oxidation of arachidonic acid carried out by PGHS.

3. The TLC plates are preconditioned in acetone prior to their use for chromatography. A glass tank is prepared by putting a sheet of filter paper (Bio-Rad) and 100 ml of acetone in it. The plates are put inside until the solvent front migrates to the top of the plates. This step lasts around 1 h. The plates are then removed from the tank and laid in a chemical hood until they appear completely dry before loading the organic extract.

4. In order to avoid subjecting the enzyme to local high concentrations of ethanol, it is important to add the solution of inhibitor last, after the enzyme has been diluted in the buffer.

5. The reaction time of 8 s has been established as being the time in the given conditions of reaction where the enzymatic reaction still approximates the linear phase and when no more than 40% of the initial substrate has been oxidized.

6. The terminating solution extracts both the residual arachidonic acid and the more polar products of oxidation in an equivalent manner. For this reason, it is not necessary to collect the entire organic phase during the extraction. Care should be taken not to pipette *ANY* water at this stage, as it would interfere with the migration of the products during the chromatography.

7. The height of solvent in the tank should not be more than 2–3 mm. To avoid diffusion of the material loaded on the plate into the solvent system, it is important that nothing be loaded at less than 5 mm (0.5 cm) from the bottom of the plate. Therefore, all material migrates upward by capillarity during the chromatography.

8. A sheet of filter paper is placed against the wall of the tank to allow saturation of the atmosphere inside the tank with the solvent system.

9. The volume of solvent added in the tank will depend on its size. The volume of solvent added is such that it reaches a depth of about 2 mm. The remaining solvent can be stored at 4°C in the glass bottle for a couple of weeks.

10. Because the product of the enzymatic reaction, PGH_2, is unstable, it rearranges readily to several other prostanoids with different polarity. For this reason and because all products are derived from PGHS activity, the entirety of the polar oxidation products migrating between the loading zone and the residual arachidonic acid are integrated in one region. The residual arachidonic acid is also integrated and the results are expressed as percent of total arachidonic acid (see Fig. 4).

11. To compare the rates of acetylation of the two PGHS isoforms, of any splice variants or of mutant enzymes by aspirin, care should be taken to assess first the specific activities of the two enzymatic preparations used. It is important to use the

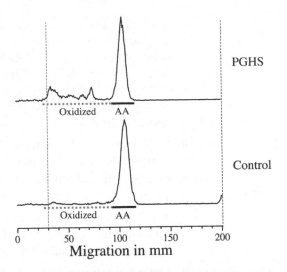

Fig. 4. Integration of radioactive regions on TLC plates. The PGHS assay is performed as described in Subheading 3.1 of this chapter. The radioactive regions of interest are determined as products of oxygenation of arachidonic acid by the enzyme (oxidized) and residual arachidonic acid (AA) for the enzymatic reaction (PGHS) and for the control reaction (Control). The dashed line represents the region to be integrated for oxidized AA and the solid line represents the region to be integrated for AA that is not oxidized. The percent of oxidized fatty acids is determined in the control reaction and subtracted from the percent of oxidized fatty acid determined in the oxidation reactions. The difference represents the percent of oxidized arachidonic acid oxidized by PGHS.

amounts of the enzymes that have equal catalytic activity in the experiments, as only enzyme molecules that are catalytically competent can be acetylated (6). This is important because the measurement of enzyme concentration by molarity (based on weight) does not account for the fact that a portion of the enzyme preparation may be catalytically inactive. For this protocol, we used PGHS isoforms with specific activities of 202 mol arachidonic acid oxidized/min/mol of PGHS-1 and 109 mol arachidonic acid oxidized/min/mol of PGHS-2. We determined the catalytic unit as the amount of PGHS that catalyzes the oxygenation of 1 nmol of arachidonic acid/min. The enzymes are reconstituted to have equal enzymatic activity, which corresponds for our enzyme preparations to final concentrations of 1.5 µM for PGHS-1 and 2.7 µM for PGHS-2.

12. The initial rate of acetylation of PGHS by aspirin is extremely fast and the time necessary for adding the 137.5 µl of enzyme, mixing and taking 20 µl is enough to lead to acetylation of part of the enzyme. In order to have a "true" zero minute time point, 18.4 µl of enzyme mix is added to 5 µl of SDS-PAGE loading buffer. Then 1 µl of [^{14}C] aspirin is added and the tube is immediately incubated at 70°C for 10 min. In these conditions, the enzyme is denatured before aspirin is added and the radioactivity associated to the enzyme represents

the nonspecific binding that may occur during the termination of the reaction. This value is subtracted from the values calculated at each time point.

13. The beads are kept in a solution of 50% ethanol in water that needs to be removed before being used. The beads need to be resuspended by gently inverting the bottle trying not to shake them as they may break generating "fines" (small debris) that could clog the column and interfere with the elution of the platelets.

14. Sepharose beads can sustain bacterial and fungus growth if kept in storage for a long time. To prevent this, the washed beads are kept in a buffer containing 0.02% sodium azide. This chemical is removed when the column is poured and washed before utilization.

15. In these conditions, the volume of packed beads is about 5.5 ml, which represents a height of ~7 cm. These chromatographic parameters are essential for a good separation of the platelets from contaminants in plasma.

16. Puncture of the vein with the needle causes local activation of platelets that can modify platelet function in the sample. For this reason, it is necessary to discard the first 3 ml of blood that is drawn.

17. Platelets are activated by glass surfaces. Therefore, no glass is used at any time during the procedure.

18. Platelets are activated when placed in a cold environment. Therefore, all steps are performed at room temperature until the inhibition experiments are done at 37°C. Care is taken to wash the platelets and perform the inhibition within 4 h following blood draw. After this time, platelet function is altered, which could confuse the interpretation of the results.

19. The white blood cells (WBC) are present at the interface between the PRP (yellow upper layer) and the red blood cells (red lower layer) in the form of a very thin white layer. It is important NOT to take any WBC as they contain PGHS and can confound the results. To avoid that, the pipette used to collect the PRP is not lowered closer than 5 mm from the interface and care is taken not to aspirate the PRP too fast as it would disrupt the layers and increase the risk of aspirating WBC.

20. In general, acidification of the plasma necessitates the addition of 80 μl of citric acid per ml of plasma. The pH should be tested with pH paper as different volunteers have different buffer capacities. Citric acid is added with a pipetman while swirling the pipette tip from the bottom to the top of the tube. The plasma can be mixed further by gently inverting the capped tube once or twice. Care should be taken not to shake the tube, as it would activate the platelets at this time. After acidification,

the PRP is kept at room temperature for 10 min before proceeding to the next step of the protocol. This time is necessary for the citric acid to achieve maximal effect.

21. This chromatographic step is necessary to remove all proteins and salt present in the plasma contaminating the pelleted platelets. The volume in which the platelets are resuspended must not be larger than 500 µl. If too large, the platelets will not be well separated from the plasma components with the volume of sepharose present in the column. If larger volumes of sepharose are used, the cells will be too dilute and will not be usable for the inhibition experiments. In these experimental conditions, the platelets are eluted in a total volume of about 1 ml.

22. The platelets are collected from the time the eluate becomes milky until the eluate becomes clear. The time has to be estimated by the person performing the chromatography. The principle is that the longer you wait, the more dilute is the resulting platelet-containing eluate, and the more you risk collecting contaminants from the plasma, which elutes after the platelets. If you stop collecting too early, you waste platelets. By following the protocol described here, the platelets are usually eluted in a total volume of about 1 ml ± 20%.

23. As for the purified enzymes, the amount of hydroperoxides generated by adding arachidonic acid to the cells attenuates the potency of the inhibitors. For this reason, the experiments are performed in the presence of a final concentration of arachidonic acid of only 2 µM.

24. The samples can be kept frozen at –20°C at this point of the protocol *BEFORE* evaporating the solvents.

25. Gas chromatography mass spectrometry is an extremely sensitive method that can detect femtomoles of contaminants. Also, yields of several steps of the derivatization process will be affected by the impurities present in the glassware. For these two reasons, the reactivials must be extensively washed according to the following protocol. The vials are soaked overnight in soapy water (*do not leave them for days*). They are then brushed with a tube brush ensuring to reach the bottom of the vial. The vials are then rinsed three times with tap water, three times with distilled water, and finally three times with ethanol. The vials are inverted in a rack and allowed to dry in air. The lids are washed separately. They are soaked in soapy water for a couple of hours, swirled for a couple of minutes and taken out (Lids are *NOT* soaked overnight with the vials as the teflon lining would come off of the lid). They are then rinsed three times with tap water, three times with distilled water, and finally three times with ethanol.

26. The esterification step is sensitive to the presence of water. Care should be taken to avoid the presence of any water in the reaction. The dichloromethane should be thoroughly dried,

and the presence of any liquid residue on the side of the tube should be checked before adding the DIPEA. If water residue is present, add 50 µl of anhydrous acetonitrile, which helps remove water by forming the azeotrope. DIPEA is anhydrous and PFB-Br should be prepared in anhydrous acetonitrile and kept for no more than a week.

27. The yield of the reaction is sensitive to the incubation time. Care should be taken not to incubate the reaction longer than 30 min.

28. Silica particles are toxic and should not be breathed. The plates should be scraped under a fume hood and a mask should be worn during the process.

29. The silation reaction is *EXTREMELY* sensitive to the presence of water. Care should be taken to avoid the presence of any water in the reaction. The ethyl acetate should be thoroughly dried, and the presence of any liquid residue on the side of the tube should be checked before adding the pyridine. Pyridine is kept in a small container on calcium hydride to trap any water dissolved from the ambient air. TMCS-BSTFA comes from a sealed ampoule opened right before the reaction.

30. Poor yields of the derivatization process can occur with problems in this step. The presence of trace amounts of water prevents the silation reaction to occur. To minimize the risks, the reaction of silation is done on the evening before or on the same morning of the day the samples are injected in the mass spectrometer. If the reaction fails, the undecane can be evaporated to dryness under nitrogen stream and the silation reaction can be done again following steps 16 and 17 of the protocol.

31. The GC/MS vials should be prepared in advance and the undecane should be transferred immediately after solubilizing the residues because the undecane will dissolve the plasticizers from the tube, which will interfere with the mass spectrometric analysis.

32. The purpose of this chapter is not to describe the method of gas chromatography mass spectrometry but the enzymatic conditions required to assay PGHS activity. In short, the derivatives will be analyzed by gas chromatography/electron capture negative chemical ionization mass spectrometry, using a DB 1701 column (15 m), with a temperature gradient from 190°C to 300°C at 20°C/min. The ions corresponding to the derivatized prostanoids are monitored by selected ion monitoring. The signals for TxB$_2$ is $m/z = 614$. The signal for the internal standard [^{2}H$_4$] TxB$_2$ is $m/z = 618$. To account for the deuterium–protium exchange at the position C12 of [^{2}H$_8$] PGE$_2$, the signals obtained at $m/z = 530$, $m/z = 531$, and $m/z = 532$ are summed (7). The signal for the internal standard [^{2}H$_4$] PGE$_2$ is $m/z = 528$.

33. Acetaminophen efficacy in some cell lines is low, and it is necessary to use a maximal concentration of the drug in the millimolar range. For this reason, a stock solution of 1 M is prepared in ethanol. The solution must be sonicated in a water bath sonicator until complete solubilization of acetaminophen.

34. Cells expressing PGHS-2 are constitutively producing prostaglandins. The purpose of the assay is to assess the PGHS activity at a given time after treatment. Therefore, we need to be able to discriminate between the prostaglandins accumulated in the cells before and during the treatment with the inhibitor from the prostaglandins synthesized at the selected time of inhibition. For this purpose, the reaction is initiated by adding deuterated arachidonic acid. Because the product of the reaction of deuterated arachidonic acid will have a higher molecular weight than the prostaglandins produced from endogenous stores of arachidonic acid, they can be differentiated by mass spectrometry.

35. The acidification is necessary to protonate the carboxylic group of prostanoids facilitating their extraction into organic solvents and/or their adsorption on the C18 cartridges.

36. The wash with heptane is necessary to remove nonpolar fatty acids that would otherwise be eluted with the prostanoids at the next step.

Acknowledgments

The authors would like to thank Taneem Amin for his technical assistance and Manju Bala and Bharati Kakkad for their critical evaluation of the manuscript.

References

1. Boutaud O, Aronoff DM, Richardson JH, Marnett LJ, Oates JA (2002) Determinants of the cellular specificity of acetaminophen as an inhibitor of prostaglandin H2 synthases. Proc Natl Acad Sci USA 99(10):7130–7135

2. Ouellet M, Percival MD (2001) Mechanism of acetaminophen inhibition of cycloxygenase isoforms. Arch Biochem Biophys 387(2): 273–280

3. Schildknecht S, Daiber A, Ghisla S, Cohen RA, Bachschmid MM (2008) Acetaminophen inhibits prostanoid synthesis by scavenging the PGHS-activator peroxynitrite. FASEB J 22(1): 215–224

4. Bala M, Chin CN, Logan AT et al (2008) Acetylation of prostaglandin H2 synthases by aspirin is inhibited by redox cycling of the peroxidase. Biochem Pharmacol 75(7):1472–1481

5. Kirkland SC (1985) Dome formation by a human colonic adenocarcinoma cell line (HCA-7). Cancer Res 45(8):3790–3795

6. Chen YN, Marnett LJ (1989) Heme prosthetic group required for acetylation of prostaglandin H synthase by aspirin. FASEB J 3(11): 2294–2297

7. Roberts LJ 2nd, Lewis RA, Oates JA, Austen KF (1979) Prostaglandin thromboxane, and 12-hydroxy-5, 8, 10, 14-eicosatetraenoic acid production by ionophore-stimulated rat serosal mast cells. Biochim Biophys Acta 575(2): 185–192

Chapter 8

Different Methods for Testing Potential Cyclooxygenase-1 and Cyclooxygenase-2 Inhibitors

Stefan Laufer and Sabine Luik

Abstract

The need for the development of selective agents, which only inhibit the mainly "harmful" cyclooxygenase-2 (COX-2) while leaving physiological COX-1 mostly unaffected, still remains, especially after the recent issues related to cardiovascular toxicity caused by some COX-2 selective agents. Thus there is still a demand for sensitive and rapid methods to assay for COX-2 selective agents. Among several in vitro testing systems the whole blood assay (WBA) is a well-known method to examine non-steroidal anti-inflammatory drugs (NSAIDs) in view of their potency to inhibit COX activity. This assay has some major advantages over enzyme-based or isolated cell assays. Emergence of artifacts due to cell separation steps is kept to a minimum and substances, even in disproportional high concentrations, can be examined outside the body in a physiological environment resembling most closely the in vivo conditions in living humans, i.e., 37°C, homeostasis, presence of all blood compounds and cell–cell interactions remain intact. While COX-1 human whole blood assays are performed within less than 2 h, for established COX-2 assays one still has to allow for an overnight incubation step before gaining the desired plasma. The aim of the assay described in this chapter is to characterize an optimized human whole blood assay (hWBA). We present a simple, fast and reliable method to examine the capacity of NSAIDs at inhibiting COX-2 activity that can be applied for rapid and routine screening purposes.

Key words: Human whole blood assay, Cyclooxygenase, Nonsteroidal antiinflammatory drugs, Drug screening

1. Introduction

Cyclooxygenase (COX) enzymes exist in two isoforms, COX-1 and COX-2. The first is expressed constitutively in many tissues, i.e., kidney (1), thrombozytes (2), and vascular endothelium (3), and contributes to the healthy state of the human body. COX-2 is the inducible isoenzyme, which is upregulated in response to cytokines, growth factors, and toxins, and displays an indicator of illnesses such as inflammation and cancer and is controversially

Samir S. Ayoub et al. (eds.), Cyclooxygenases: Methods and Protocols, Methods in Molecular Biology, vol. 644, DOI 10.1007/978-1-59745-364-6_8, © Springer Science+Business Media, LLC 2010

discussed in connection with Alzheimer's disease (4–6). The class of nonsteroidal antiinflammatory drugs (NSAIDs) is widely distributed first choice agents for the treatment of rheumatism and other inflammatory diseases. The development of selective agents that inhibit the "bad" COX-2 contemporaneously leaving the "good" COX-1 fairly untouched is a very sophisticated objective in drug research.

First in vitro testing systems that investigate the activity of COX inhibitors were described at the beginning of the 1980s, i.e., by Patrono and coworkers (7) but most assays were developed about 10 years later. By improving these testing systems, which are based on the use of cell lines cultured in the presence of the test compound (COS-1 cells (8), CHO-cells (9), RBL-1 cells (10)), new assays came along that used isolated enzyme preparations and isolated human blood cells (9, 11). When taking isolated enzymes, one avoids disturbing endogenous influences and receives exact information about the actual and sole impact of the drug on the enzyme. On the other hand, these assays do not resemble pathophysiological conditions and results can hardly be conferred to the human organism because important parameters such as cell–cell interactions, plasma binding of the drug, and other blood compounds are neglected.

Isolated cells do have the advantage of being derived from fresh whole blood. But apart from the advantage of a minimized protein content, this system deals with the same problems as the isolated enzyme assay, namely that other interactions besides the drug–cell interplay are being ignored. Nevertheless, enzyme and cell-based assays are useful and powerful accompanying tools to determine preliminary drug kinetics and dose–response relationships.

However, all assays that are based on the separation of cells from fresh blood have in common that they are vulnerable to artifacts, time-consuming and cost-intensive and are therefore inappropriate for daily and routine testing. With this knowledge, other test systems that use human whole blood were developed precociously (10, 12–14) in addition to the enzyme and isolated cell assays. Using whole blood minimizes the introduction of artifacts that would occur during cell separation steps; in addition all cell–cell interactions persist, and a physiological environment, including plasma protein levels, that resembles closely the in vivo situation is provided.

In this chapter, we describe an ameliorated simple assay for testing COX-2 inhibitors in heparinized human whole blood challenged with the bacterial endotoxin lipopolysaccharide (LPS). Enzyme activity is subsequently measured by determining prostaglandin E_2 (PGE_2) content in plasma supernatant with a nonradioactive enzyme immunoassay (EIA). Procedures and results are compared with established assays and literature data. By saving a considerable amount of time, this assay is a convenient and reliable tool for daily and routine testing of NSAIDs in human whole blood.

2. Materials

2.1. COX-1 Whole Blood Assay

2.1.1. Stimulation of Blood

1. Calciumionophore: 2.5 mg Calciumionophore is diluted in 1 ml ethanol and stored at –20°C.

2. Cremophor-EL/ethanol: 116.9 g Cremophor-EL (BASF) is mixed with 50.0 g ethanol 99% in a volumetric flask (250 ml) and stored limitless under cover of darkness. The volumetric flask must be hermetically sealed.

3. DPBS Buffer: 4.0 g NaCl, 0.1 g KCl, 0.1 g KH_2PO_4, 0.575 g Na_2HPO_4, and 0.5 g Glucose are diluted in water in a volumetric flask (500 ml) and stored at 4°C for a maximum of 7 days (see Note 2).

4. Thromboxane Synthetase Inhibitor (TXBSI) stock solution: 1.4 mg TXBSI (was synthesized in our lab) is diluted in 5 ml Cremophor-EL/ethanol and stored in single-use aliquots at –20°C.

5. Test compounds: Test compounds are solved 10 mM in Cremophor-EL/ethanol and stored in single-use aliquots at –20°C. As compounds sometimes are difficult to be dissolved, stock solutions must be put in ultrasonic bath. Working solutions must be freshly prepared by diluting the stock solution 1:10 in DPBS buffer. Final working concentration is 0.01 μM (see Note 2).

6. DPBS–Gentamicin buffer: 12.5 mg Gentamicin is diluted in 50 ml DPBS buffer (see Note 3).

7. Thromboxane Synthetase Inhibitor (TXBSI): Approximately, 4 ml DPBS–Gentamicin buffer is added to a volumetric flask (5 ml) and 0.5 ml of the TXBSI stock solution is mixed with the DPBS–Gentamicin buffer (see Note 4).

8. Accordant amount of heparinized blood of two donors (see Note 5).

9. Dilution of the test compounds:

1,000 μM	→ 900 μl DPBS buffer + 100 μl of	10,000 μM Stock test compound solution
100 μM	→ 900 μl DPBS buffer + 100 μl of	1,000 μM Test compound
10 μM	→ 900 μl DPBS buffer + 100 μl of	100 μM Test compound
1 μM	→ 900 μl DPBS-buffer + 100 μl of	10 μM Test compound

(see Note 4)

Test compounds are diluted further when added to blood to give final working concentrations of 10–0.01 µM.

2.1.2. PGE$_2$-ELISA (15)

PGE$_2$ was measured using a commercial ELISA kit from Assay Designs Inc., Ann Arbor, USA. Below is a list of the kit's components with a description of the preparation methods.

1. This assay is based on the use of a 96-wells plate coated with goat anti-Mouse IgG antibody. The plates are made from break-apart strips coated with goat antibody specific to mouse IgG (see Note 6).

2. Alkaline Phosphatase PGE$_2$ Conjugate: A blue solution of alkaline phosphatase conjugated with PGE$_2$. Ready to use.

3. PGE$_2$, EIA Antibody: A yellow solution of a monoclonal antibody to PGE$_2$. Ready to use.

4. Assay Buffer Concentrate: Tris buffered saline containing proteins and sodium azide as preservative. Prepare the Assay buffer by diluting 10 ml of the concentrate with 90 ml of bidistillated water. The dilution can be stored at room temperature until the kit expiration date, or for 3 months, whichever is earlier.

5. Wash Buffer Concentrate: Tris buffer saline containing detergents. Prepare the Wash buffer by diluting 5 ml of the supplied concentrate with 95 ml of bidistillated water. This can be stored at room temperature until the kit expiration date, or for 3 months, whichever is earlier.

6. Prostaglandin E$_2$ Standard: A solution of 50.000 pg/ml PGE$_2$ (see Note 7).

7. pNpp Substrate: A solution of *p*-nitrophenyl phosphate in buffer. Ready to use.

8. Stop Solution: A solution of trisodium phosphate in water. Keep tightly capped.

9. PGE$_2$ Assay Layout Sheet and Plate Sealers.

2.2. COX-2 Whole Blood Assay

2.2.1. Stimulation of Blood

1. DPBS Buffer: (see Subheading 2.1.1).

2. Cremophor-EL/ethanol: (see Subheading 2.1.1).

3. Thromboxane Synthetase Inhibitor (TXBSI) stock solution: (see Subheading 2.1.1).

4. DPBS–Gentamicin Buffer: (see Subheading 2.1.1).

5. Thromboxane Synthetase Inhibitor (TXBSI): (see Subheading 2.1.1).

6. Acetylsalicylic Acid Solution: 2.0 mg Acetylsalicylic acid is diluted in 10.0 ml DPBS–Gentamicin. Working concentration is defined as 10 µg/ml.

7. LPS-Solution: 1.0 mg LPS (SIGMA) is diluted in 5 ml DPBS–Gentamicin. Working concentration is defined as 10 µg/ml.

8. 40 ml heparinized blood of two donors (see Note 5).

9. Samples: Initial drug concentration range from 10 µM to 10 pM and is prepared by diluting the stock solution in DPBS–Gentamicin. Starting from 10 mM, 1:1,000-fold dilution is required to obtain 10 µM.

2.2.2. PGE₂-ELISA (15)

(see 2.1.2.)

2.3. COX-1 and 5-Lipoxygenase (5-LO) in Peripheral Blood Mononuclear Cell (PBMC)

2.3.1. Isolation and Washing of Platelets and Granulocytes

1. Percoll solution (100%): 1.753 g Sodium chloride is solved in 200 ml Percoll. The solution can be stored at 4°C up to 2 months.

2. Percoll solution (74%): 74 g 100% Percoll solution is mixed with 26 g 0.9% Sodium chloride solution. The solution can be stored at 4°C up to 2 months.

3. Percoll solution (55%): 55 g 100% Percoll solution is mixed with 45 g 0.9% Sodium chloride solution. The solution can be stored at 4°C up to 2 months.

4. Percoll solution (44%): 44 g 100% Percoll solution is mixed with 56 g 0.9% Sodium chloride solution. The solution can be stored at 4°C up to 2 months.

5. Sodium chloride solution (0.9%): 4.5 g Sodium chloride is solved in 500 ml water (see Note 1).

6. Lysis medium: 4.15 g Ammonium chloride, 0.019 g Sodium EDTA, and 0.5 g Kalium hydrogen carbonate are dissolved in 500 ml water. The solution can be stored at 4°C up to 2 weeks (see Note 1).

7. DPBS-Buffer: (see 2.1.1.)

8. EDTA Solution: 1.435 g sodium EDTA and 0.105 g sodium chloride are diluted in 50 ml water. The solution can be stored at 4°C up to 2 weeks (see Note 1).

9. Calcium chloride solution: 0.111 g calcium chloride and 0.8 g Sodium chloride are solved in 100 ml water (see Note 1).

10. NDGA solution (Nordihydroguaiaretic acid): 1.5 mg NDGA is dissolved in 250 ml water. Put the volumetric flask (250 ml) into the ultrasonic bath. All solutions must be brought at room temperature before use (see Note 1).

11. EDTA/Macrophages-SFM Solution: add 7.8 ml EDTA solution to a 200 ml volumetric flask and fill up with macrophages-SFM solution.

12. Heparinized blood of four donors (see Note 5).

2.3.2. LTB₄-ELISA (16)

1. Goat anti-Rabbit IgG Microtiter Plate, 96-Wells (Assay Designs, Inc., Ann Arbor, USA) A plate using break-apart strips coated with goat antibody specific to rabbit IgG (see Note 8).

2. LTB$_4$ EIA conjugate: A blue solution of alkaline phosphatase conjugated with LTB$_4$ (see Note 9).

3. LTB$_4$ EIA antibody: A yellow solution of a rabbit polyclonal antibody to LTB$_4$.

4. Assay buffer concentrate: Tris buffered saline, containing proteins, detergents, and sodium azide as preservative. Prepare the Assay buffer by diluting 10 ml of the concentrate with 90 ml water. This can be stored at room temperature until the expiration date, or for 3 months, whichever is earlier (see Note 1).

5. Wash buffer concentrate: Tris buffered saline containing detergents. Prepare the wash buffer by diluting 5 ml of the concentrate with 95 ml water. This can be stored at room temperature until the kit expiration date, or for 3 months, whichever is earlier (see Note 1).

6. Leukotriene B$_4$ Standard: A solution of 120.000pg/ml LTB$_4$.

7. pNpp Substrate: A solution of *p*-nitrophenylphosphate in buffer. Ready to use.

8. Stop solution: A solution of trisodium phosphate in water. Keep tightly capped.

9. LTB$_4$ Assay Layout Sheet.

10. Plate Sealer.

2.3.3. PGE₂-ELISA (15)

(see Subheading 2.1.2.)

3. Methods

3.1. COX-1 Whole Blood Assay

3.1.1. Stimulation of Blood

1. Before starting your assay, cool the centrifuge to 4°C.

2. The water bath must be brought to 37°C.

3. Start pipetting the blood of donor one and store the other on the titramax at 37°C at a low rpm to avoid accumulation of the blood cells (see Note 10).

4. Four samples can be incubated at the same time (see Note 11).

5. Pipet 500 µl of the heparinized blood into each tube.

6. Add 5 µl of the TXBSI into each tube (see Note 12).

7. Pipet 5 μl of the dilutions into the appropriate caps-keep a 15 s rhythm.

8. Pipet 5 μl of the DPBS buffer into the tubes for the basal level and the maximal stimulation data.

9. Wash the tip out in the blood suspension (see Note 13).

10. Put the tubes into the water bath for exact 15 min.

11. After 15 min incubation time, pipet 2.5 μl of the Calciumionophor dilution in all tubes for starting the reaction (except the caps for the basal level) (see Note 14).

12. Keep the 15 s rhythm.

13. Pipet 2.5 μl of the DPBS buffer into the tubes for basal level (instead of Calciumionophor dilution).

14. An overview of the single steps is shown in Table 1.

15. Incubation takes exact 1 min.

16. Take the tubes out of the water bath and store them on iced water for 5 min to stop the reaction.

17. Tubes are centrifuged at $1,000 \times g$ and 4°C for 20 min.

18. Plasma is taken out and put into a fresh tube and stored at −20°C.

19. If the PGE_2-ELISA of the compounds cannot be tested within 1 week store them at −80°C.

Table 1
(COX-1): Brief overview of the individual pipetting steps for COX-1 determination

Basal level	500 μl Blood
	5 μl TXBSI
	5 μl DPBS-buffer
	2.5 μl DPBS-buffer
Maximal stimulation	500 μl Blood
	5 μl TXBSI
	5 μl DPBS-buffer
	2.5 μl Calciumionophor-dilution
Samples	500 μl Blood
	5 μl TXBSI
	5 μl Accordant inhibitor-dilution
	2.5 μl Calciumionophor-dilution

20. Handle remaining blood and blood of donor two in the exact same manner (see Note 15).

3.1.2. PGE₂-ELISA

1. Refer to the Assay Layout Sheet to determine the number of wells to be used and put any remaining wells with the desiccant back into the pouch and seal the Ziploc. Store unused wells at 4°C (see Table 2).

2. Each value should be measured in double determination.

3. Preparation of the standards according to the manufacturer's instruction (see Table 3) (see Note 16).

4. Preparation of the samples:

   380 µl Assay buffer is mixed with 20 µl of the corresponding samples. Remaining sample volume is stored at –20°C (see Note 17).

Table 2
(COX-1/COX-2/COX-1 and 5-LO PBMC): Possible pipetting scheme

St. 1	5,000 [pg/ml]	Blank	Blank	Stim2	Stim2	S	S	S	S	Stim3	Stim3
				I	I	I	I	II	II	II	II
St. 2	2,500 [pg/ml]	B0	B0	S	S	S	S	S	S	S	S
				I	I	I	I	II	II	II	II
St. 3	1,250 [pg/ml]	TA	TA	S	S	S	S	S	S	S	S
				I	I	I	I	II	II	II	II
St. 4	625 [pg/ml]	NSB	NSB	S	S	S	S	Stim2	Stim2	S	S
				I	I	I	I	II	II	II	II
St. 5	312.5 [pg/ml]	Basal	Basal	S	S	Stim4	Stim4	S	S	S	S
		I	I	I	I	I	I	II	II	II	II
St. 6	156.25 [pg/ml]	Stim1	Stim1	S	S	Basal	Basal	S	S	Stim4	Stim4
		I	I	I	I	II	II	II	II	II	II
St. 7	78.125 [pg/ml]	S	S	Stim3	Stim3	Stim1	Stim1	S	S		
		I	I	I	I	II	II	II	II		
St. 8	39.1 [pg/ml]	S	S	S	S	S	S	S	S		
		I	I	I	I	II	II	II	II		

I donor 1, *II* donor 2, *S* sample [µM], *TA* Total activity, *NSB* Non-specific binding, *Stim* maximal stimulation, *Basal* basal level

Typical quality control parameters:

$TA = 2.815 \times 10 = 28.15$

$\%NSB = 0.02\%$

$\%B0/TA = 2.25\%$

Quality of fit = 1.00 (calculated from four parameter logistic curve fit)

Table 3
(COX-1): Brief overview of the various dilution steps
of the standard

Standard number	Final concentration [pg/ml]	Standard [µl]	Assay buffer [µl]	Dilution
1	5,000	100	900	1:10
2	2,500	500 of St. 1	500	1:2
3	1,250	500 of St. 2	500	1:2
4	625	500 of St. 3	500	1:2
5	313	500 of St. 4	500	1:2
6	156	500 of St. 5	500	1:2
7	78.1	500 of St. 6	500	1:2
8	39.1	500 of St. 7	500	1:2

5. Pipet 100 µl Assay buffer into the NSB and B0 (0 pg/ml Standard) wells.

6. Add 100 µl of standards 1 through 8 into the appropriate wells and 100 µl of the samples into the corresponding wells (see Note 18).

7. 50 µl of the Assay buffer must be added into the NSB wells.

8. Pipet 50 µl of conjugate into each well, except the Blank and TA wells.

9. Add 50 µl of the antibody solution into each well, except the Blank, TA, and NSB wells (see Note 19).

10. Incubate the plate at room temperature on a plate shaker for 120 min at ~500 rpm. The plate may be covered with the plate sealer provided, if so desired.

11. Empty the contents of the wells and wash by adding 400 µl of wash solution to every well. Repeat the wash two more times for a total of three washes.

12. After the final wash, empty the wells and tap the plate dry on a lint free paper towel.

13. Add 5 µl of the blue conjugate to the TA wells.

14. Add 200 µl of the pNpp substrate solution to every well. Incubate at room temperature for 45 min without shaking.

15. Pipet 50 µl of stop solution to every well (see Note 20).

16. Blank the plate reader against the Blank wells and read the optical density at 405 nm, preferably with correction between

570 and 590 nm. If the plate reader is not able to be blanked against the Blank wells, manually subtract the mean optical density of the Blank wells from all readings.

3.1.3. Calculation of PGE₂ Concentration

1. The average of the PGE_2 concentration of stimulation data must be calculated first.

2. The average of PGE_2 concentration of basal data has to be subtracted from the average of each sample (including the values for maximal stimulation).

3. To generate the percentage inhibition, use the following formula

$$\text{Inhibition } [\%] = 100 - (C_{sample}/C_{stim}) \times 100.$$

$C = PGE_2$ concentration.

4. Using Logit-Log paper plot percent inhibition versus concentration of PGE_2 for the standards. Approximate a straight line through the points. The concentration of PGE_2 in the unknowns can be determined by interpolation. Otherwise, concentration of PGE_2 can be measured using a special software.

5. To get the IC_{50} data Logit-Log paper plot, Percent Inhibition versus concentration of the accordant sample is used.

6. TA and NSB values are quality control parameters (see Note 21).

3.2. COX-2 Whole Blood Assay

3.2.1. Stimulation of Blood

1. Two 20 ml samples of fresh heparinized (15 units/ml) venous blood per donor are pooled, aliquoted (800 μl/tube), and mixed with 10 μl TXBSI per tube (see Note 13).

2. Start pipetting blood of donor one and store the other on the titramax at a low rpm to avoid accumulation of the blood cells (see Note 10).

3. Refer to the Assay Layout Sheet to determine the number of caps to be used.

4. Acetylsalicylic acid (10 μg/ml) is added at time zero to exclude any contribution of COX-1 to metabolite formation (see Note 22).

5. After adding test substances or DPBS (for maximal stimulation and basal data), samples are equilibrated in a humidified incubator (37°C, 5% CO_2) for 15 min.

6. Pipet LPS (10 μg/ml) in the corresponding wells to induce COX-2 expression.

7. 50 μl DPBS–Gentamicin buffer is added into the tubes for basal data instead of LPS.

8. Samples are incubated for 5 h in the humidified incubator. An overview of the single steps is shown in Table 4.

9. Handle blood of donor two in the exact same manner.

Table 4
(COX-2): brief overview of the individual pipetting steps for COX-2 determination

Basal level	800 µl Blood
	10 µl TXBSI
	50 µl Acetylsalicylic acid
	100 µl DPBS–Gentamicin-buffer
	50 µl DPBS–Gentamicin-buffer
Maximal stimulation	800 µl Blood
	10 µl TXBSI
	50 µl Acetylsalicylic acid
	100 µl DPBS-Gentamicin-buffer
	50 µl LPS-Dilution
Samples	800 µl Blood
	10 µl TXBSI
	50 µl Acetylsalicylic acid
	100 µl Accordant inhibitor dilution
	50 µl LPS-dilution

10. Reaction is stopped by adding an equal volume (=1,000 µl) of iced DPBS buffer into the aliquots and further cooling on ice for 10 min.

11. Samples are centrifuged ($1,000 \times g$, 4°C, 15 min) and plasma is removed and stored at −20°C.

12. PGE_2 concentration in thawed plasma supernatants is determined by correlate enzyme immunoassay as an indicator of COX-2 activity according to the manufacturers' instructions.

3.2.2. PGE_2-ELISA (15)

(see Subheading 3.2.)
(see Table 2)
(see Table 5)

3.2.3. Calculation of PGE_2 Concentration

(see Subheading 3.1.3.)

3.3. COX-1 and 5-LO PBMC

3.3.1. Isolation and Washing of Platelets and Granulocytes

1. Four 20 ml samples of fresh heparinizied (15 units/ml) venous blood per donor are pooled. Start pipetting blood of donor one and two and store the other two on the titramax at a low rpm to avoid accumulation of the blood cells.

2. Four falcon tubes per donor are necessary.

Table 5
(COX-2): brief overview of the various dilution steps of the standard

Standard number	Final concentration [pg/ml]	Standard [μl]	Assay buffer [μl]	Dilution
1	5,000	100	900	1:10
2	2,500	500 of St. 1	500	1:2
3	1,250	500 of St. 2	500	1:2
4	625	500 of St. 3	500	1:2
5	313	500 of St. 4	500	1:2
6	156	500 of St. 5	500	1:2
7	78.1	500 of St. 6	500	1:2
8	39.1	500 of St. 7	500	1:2

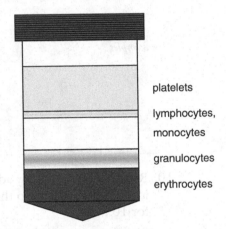

platelets

lymphocytes, monocytes

granulocytes

erythrocytes

Fig. 1. Different cell layers formed after centrifugation.

3. Pipet 3 ml Percoll solution (74%) in a falcon tube made of polystyrene and put a thin polystyrene platelet on the top of the solution (see Note 23).

4. Pipet 3 ml Percoll solution (55%) into the same falcon tube. Handle carefully to avoid mixing of the two different solutions (see Note 24).

5. Pipet 2 ml Percoll solution (44%) into the same falcon tube. Try to avoid mixing of the different solutions again.

6. Pipet 5 ml of the pooled blood into the accordant falcon tubes. Use tweezers to take the polystyrene platelet out of the falcon tube. Close the falcon tube and centrifuge the tubes for 20 min ($350 \times g$, 20°C).

7. After centrifugation, different layers should be apparent in the tube (see Fig. 1).

8. Take a plastic pasteurpipet and pipet the layer, including the platelets into a fresh centrifugation tube. Two falcon tubes should be pooled.

9. The eight tubes are centrifuged at $1,000 \times g$ and 20°C for 20 min (see Note 25).

10. During the 20 min, the granulocytes can be isolated.

11. To isolate the granulocytes, take new pasteurpipets (a fresh one for each donor) and put the layer including the granulocytes in new falcon tube.

12. The 16 tubes are centrifuged at $350 \times g$ and 20°C for 10 min.

13. During this time, supernatant of the platelets must be removed (until the 1.5 ml mark) (see Note 26).

14. Pipet 3 ml of the EDTA/Macrophages-SFM solution and resuspend the cell pellet.

15. Platelet suspension is centrifuged at $1,000 \times g$ and 20°C for 20 min.

16. Repeat this step twice.

17. Final step is (after removing the supernatant) to pipet 1 ml of the EDTA/Macrophages-SFM solution into all falcons.

18. Resuspend the pellets and pool them (see Note 27).

19. Take the falcon tubes including the granulocytes pellets, remove the supernatant, and fill the tubes until the 3 ml mark with DPBS buffer.

20. Centrifuge the granulocytes suspension at $150 \times g$ and 20°C for 10 min.

21. Remove the supernatant and pipet 0.5 ml DPBS buffer in each tube for resuspending the pellet.

3.3.2. Determination of the Cell Count

3.3.2.1. Granulocytes

1. Put 1,000 µl Trypan blue solution into a tube (see Note 28).

2. Add 100 µl granulocyte suspension into the same cap and mix it.

3. An incubation time of 5 min is necessary (see Note 29).

4. Put the cover slip on the Abbe-Zeiss counting cell chamber in a way that Newton's rings are visible.

5. Switch the microscope on and put the Abbe-Zeiss counting cell chamber on the microscope.

6. After 4 min incubation time, pipet 10 µl cell suspension under the cover slip (see Note 30).

7. Scan the cross and count the cells (see Note 31).

8. Clean the chamber and cover slip and repeat the counting once.

3.3.2.2. Platelets

1. Pipet $200\,\mu l$ EDTA/Macrophages-SFM-solution into the same cap (see Note 28).

2. Add $10\,\mu l$ platelet suspension into the same cap.

3. An incubation time of 3 min is necessary.

4. Put the cover slip on the Abbe-Zeiss counting cell chamber in a way that Newton's rings are visible.

5. Switch the microscope on and put the Abbe-Zeiss counting cell chamber on the microscope.

6. Add $10\,\mu l$ cell suspension under the cover slip and count the cells.

3.3.3. Calculation of the Cell Count

To calculate the number of cells per ml, uses the following formulas:

3.3.3.1. Granulocytes

$$C/ml = \frac{\sum counted - cells \times 10 \times 16 \times 10^4}{20}$$

Volume (Abbe-Zeiss counting cell chamber) $= 10^4$
Number of squares $= 16$
Dilution factor $= 10$
Squares counted $= 20$

3.3.3.2. Platelets

$$C/ml = \frac{\sum counted - cells \times 20 \times 256 \times 10^4}{25}$$

Volume (Abbe-Zeiss counting cell chamber) $= 10^4$
Number of squares $= 256$
Dilution factor $= 20$
Squares counted $= 25$

3.3.4. Setting of the Cell Count

Final concentration of granulocytes is termed 1.25×10^6 cells/ml. To get these concentrations, use DPBS buffer for dilution.

3.3.4.1. Granulocytes

1. Calculation of the maximal volume of cell suspension including the requested cell count:

$$Vcell - suspension - max = \frac{C/ml \times V_{present - cell - suspension}}{1.25 \times 10^6 C/ml}$$

C/ml = present calculated cell count

$V_{present-cell-suspension}$ = Volume after pooling the granulocyte suspensions

$1.25 \times 10^6 C/ml$ = required cell count

2. If the calculated volume is more than 50 ml, calculate the volume for maximal 50 ml as followed:

$$Vrequired - cell - suspension = \frac{50ml \times V_{present - cell - suspension}}{V_{maximal - cell - suspension}}$$

3.3.4.2. Platelets

Final concentration of platelets is termed 0.3×10^8 cells/ml. To get these concentrations, use EDTA/Macrophages-SFM solution for dilution.

1. Calculation of the maximal volume of cell suspension including the requested cell count:

$$V_{maximal-cell-suspension} = \frac{C/ml \times V_{present-cell-suspension}}{0.3 \times 10^8 C/ml}$$

$C/ml =$ present calculated cell count
$V_{present-cell-suspension} =$ Volume after pooling the granulocyte suspensions
$0.3 \times 10^8 C/ml =$ required cell count

2. If the calculated volume is more than 50 ml, calculate the volume for maximal 50 ml as followed:

$$Vrequired - cell - suspension = \frac{50ml \times V_{present-cell-suspension}}{V_{maximal-cell-suspension}}$$

3.3.5. Stimulation of the Granulocytes

1. The granulocytes suspension should be stored the whole time in the water bath to avoid accumulation of cells (see Note 32).

2. Pipet 800 µl of the adjusted cell suspension into each tube (see Note 33).

3. Add 10 µl of the inhibitor dilutions into the corresponding tube. (see Note 12)

4. Normally, nothing has to be pipetted into the tube for the stimulation data (instead of the inhibitor dilutions). In dilution series using DMSO as solvent, the corresponding amount of DMSO can be added into the tubes for the stimulation data.

5. Compounds have to be incubated for 5 min on the shaking water bath. Use a timer.

6. After 5 min, put 200 µl Calcium chloride solution into each tube. Take the tube out of the water bath. Keep a 15 s rhythm.

7. Compounds have to be incubated another 5 min.

8. Switch the ultrasonic bath on.

9. Take the tubes out of the shaking water bath and add 10 µl of the Calciumionophor solution into each tube. Shake the tube manual. Keep the 15 s rhythm. Put the tubes back into the shaking water bath for 2 min (see Note 12).

10. A summary of all steps is given in Table 6.

11. After 2 min, take the tubes out of the water bath and put them into the iced ultrasonic bath. Add 1,000 µl of the NDGA solution into each tube. Keep the 15 s rhythm.

Table 6
COX-1 and 5-LO PBMC in granulocytes: Brief overview of the individual pipetting steps for granulocytes stimulation

Maximal stimulation	800 µl Adjusted cell suspension
	200 µl Calcium chloride solution
	10 µl Calciumionophor solution
	1,000 µl NDGA
	1,000 µl NDGA
	1,000 µl Aqua bidist.
Samples	800 µl Adjusted cell suspension
	10 µl Accordant inhibitor dilution
	200 µl Calcium chloride solution
	10 µl Calciumionophor solution
	1,000 µl NDGA
	1,000 µl NDGA
	1,000 µl Aqua bidist.

12. The compounds stay another 5 min in the iced ultrasonic bath.

13. Pipet 1,000 µl of the NDGA solution and 1,000 µl of purified water into each tube.

14. Mix the suspension and transfer 1 ml of the suspension into a new cap for centrifugation ($40,000 \times g$, 12 min, 5°C).

15. Transfer the supernatant into a new cap and store it at −20°C for further examination.

16. If the compounds are not being tested within 1 week, store them at −80°C.

17. Handle rest of the compounds in the exact same manner.

18. An overview of all steps is given in Fig. 2.

3.3.6. Stimulation of the Platelets

1. The platelet suspension should be stored the whole time on the water bath to avoid accumulation of the cells (see Note 34).

2. Pipet 800 µl of the adjusted cell suspension into each tube (see Note 33).

3. Add 10 µl of the inhibitor dilutions into the accordant tube (see Note 12).

4. Normally, nothing has to be pipetted into the tube for the stimulation data (instead of the inhibitor dilutions). In dilution

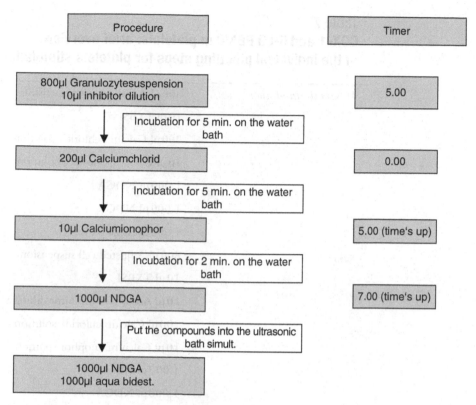

Fig. 2. Overview of granulocyte stimulation including timer.

series using DMSO as solvent, the accordant amount of DMSO can be added into the tubes for the stimulation data.

5. Compounds have to be incubated for 5 min on the shaking water bath. Use a timer!

6. After 5 min add 200 µl Calcium chloride solution into each tube. Take the tube out of the water bath. Keep a 15 s rhythm.

7. Compounds have to be incubated another 5 min.

8. Switch the ultrasonic bath on.

9. Take the tubes out of the shaking water bath and pipet 10 µl of the Calciumionophor solution into each tube. Shake the tube manual. Keep the 15 s rhythm. Put the tubes back into the shaking water bath for another 2 min (see Note 12).

10. A summary of all steps is given in Table 7.

11. After 2 min, take the tubes out of the water bath and put them into the iced ultrasonic bath. Pipet 1,000 µl of the NDGA solution into each tube. Keep the 15 s rhythm.

12. The compounds stay another 5 min in the iced ultrasonic bath.

13. Pipet 1,000 µl of the NDGA solution and 1,000 µl of purified water into each tube.

Table 7
COX-1 and 5-LO PBMC in platelets: brief overview
of the individual pipetting steps for platelets stimulation

Maximal stimulation	800 µl Adjusted cell suspension
	10 µl TXBSI
	200 µl Calcium chloride solution
	10 µl Calciumionophor solution
	1,000 µl NDGA
	1,000 µl NDGA
	1,000 µl Aqua bidist.
Samples	800 µl Adjusted cell suspension
	10 µl TXBSI
	10 µl Accordant inhibitor dilution
	200 µl Calciumchloride solution
	10 µl Calciumionophor solution
	1,000 µl NDGA
	1,000 µl NDGA
	1,000 µl Aqua bidist.
	1,000 µl Aqua bidist.

14. Mix the suspension and transfer 1 ml of the suspension into a new cap for centrifugation (40,000 × g, 12 min, 5°C).

15. Pipet the supernatant into a new cap and store it at –20°C for further examination.

16. If the compounds are not being tested within 1 week, store them at –80°C.

17. Handle rest of the compounds in the exact same manner.

18. An overview is given in Fig. 3.

3.3.7. LTB$_4$-ELISA

1. Refer to the Assay Layout Sheet to determine the number of wells to be used and put any remaining wells with the desiccant back into the pouch and seal the Ziploc. Store unused wells at 4°C (see Table 8).

2. Each value should be measured in double determination.

3. Preparation of the standards according to the manufacturer's instruction (see Table 9) (see Note 16).

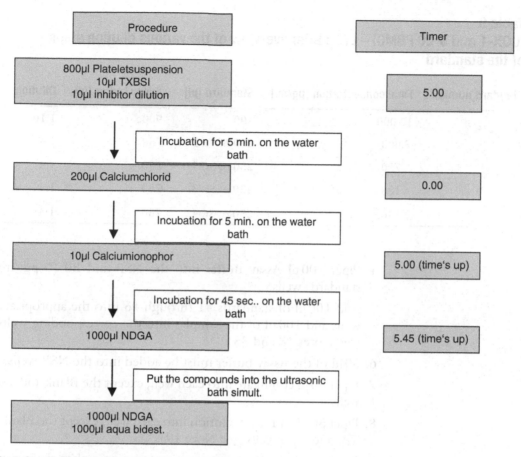

Fig. 3. Overview of platelet stimulation including timer.

Table 8
(COX-1 and 5-LO PBMC) – LTB$_4$: Possible pipetting scheme

St. 1	12,000[pg/ml]	NSB	NSB	S	S	S	S	S	S	Stim7	Stim7
St. 2	3,000 [pg/ml]	Basal	Basal	S	S	Stim1	Stim4	S	S	S	S
St. 3	750 [pg/ml]	Stim1	Stim1	S	S	S	S	S	S	S	S
St. 4	188 [pg/ml]	S	S	S	S	S	S	Stim6	Stim6	S	S
St. 5	46.9 [pg/ml]	S	S	Stim3	Stim3	S	S	S	S	S	S
Blank	Blank	S	S	S	S	S	S	S	S	Stim8	Stim8
B0	B0	S	S	S	S	Stim5	Stim5	S	S	S	S
TA	TA	Stim2	Stim2	S	S	S	S	S	S	S	S

S sample [µM], *TA* Total activity, *NSB* Nonspecific binding, *Stim* maximal stimulation, *Basal* basal level
Typical Quality Control Parameters:
$\quad$ TA = 1.466 × 10 = 14.66
$\quad$ %NSB = 0.0%
$\quad$ %B0/TA = 5.37%
$\quad$ Quality of fit = 1.00 (calculated from four parameter logistic curve fit)

Table 9
(COX-1 and 5-LO PBMC) – LTB$_4$: Brief overview of the various dilution steps of the standard

Standard number	Final concentration [pg/ml]	Standard [µl]	Assay buffer [µl]	Dilution
1	12,000	100	900	1:10
2	3,000	250 of St. 1	750	1:4
3	750	250 of St. 2	750	1:4
4	188	250 of St. 3	750	1:4
5	46.9	250 of St. 4	750	1:4

4. Pipet 100 µl Assay Buffer into the NSB and B0 (0 pg/ml standard) wells.

5. Add 100 µl of standards #1 through #5 into the appropriate wells and 100 µl of the samples into the corresponding wells (see Notes 18 and 35).

6. 50 µl of the Assay buffer must be added into the NSB wells.

7. Pipet 50 µl conjugate into each well, except the Blank and TA wells.

8. Pipet 50 µl antibody solution into each well, except the Blank, TA, and NSB wells (see Note 19).

9. Incubate the plate at room temperature on a plate shaker for 120 min at ~500 rpm.

10. Empty the contents of the wells and wash by adding 400 µl of wash solution to every well. Repeat the wash two more times for a total of three washes.

11. After the final wash, empty or aspirate the wells and firmly tap the plate on a free paper towel to remove any remaining wash buffer.

12. Pipet 5 µl of the blue conjugate to the TA wells.

13. Add 200 µl of the pNpp substrate solution to every well. Seal the plate and incubate 37°C for 120 min.

14. Pipet 50 µl of stop solution to every well (see Note 20).

15. Blank the plate reader against the Blank wells and read the optical density at 405 nm, preferably with correction between 570 and 590 nm. If the plate reader is not able to be blanked against the Blank wells, manually subtract the mean optical density of the Blank wells from all readings.

3.3.8. Calculation of LTB$_4$ Concentration   (see Subheading 3.1.3.)

3.3.9. PGE$_2$-ELISA (15)   (see Subheading 3.2.)
(see Table 10)
(see Table 11)

Table 10
(COX-1 and 5-LO PBMC) – PGE$_2$: Brief overview of the various dilution steps of the standard

Standard number	Final concentration [pg/ml]	Standard [µl]	Assay buffer [µl]	Dilution
1	5,000	100	900	1:10
2	2,500	500 of St. 1	500	1:2
3	1,250	500 of St. 2	500	1:2
4	625	500 of St. 3	500	1:2
5	313	500 of St. 4	500	1:2
6	156	500 of St. 5	500	1:2
7	78.1	500 of St. 6	500	1:2
8	39.1	500 of St. 7	500	1:2

Table 11
(COX-1 and 5-LO PBMC) – PGE$_2$: Possible pipetting scheme

St. 1	5,000 [pg/ml]	Blank	Blank	S	S	S	S	S	S	S	S
St. 2	2,500 [pg/ml]	B0	B0	S	S	S	S	Stim5	Stim5	S	S
St. 3	1,250 [pg/ml]	TA	TA	Stim2	Stim2	S	S	S	S	S	S
St. 4	625 [pg/ml]	NSB	NSB	S	S	S	S	S	S	Stim7	Stim7
St. 5	312.5 [pg/ml]	Basal	Basal	S	S	Stim4	Stim4	S	S	S	S
St. 6	156.25 [pg/ml]	Stim1	Stim1	S	S	S	S	S	S	S	S
St. 7	78.125 [pg/ml]	S	S	S	S	S	S	Stim6	Stim6	S	S
St. 8	39.1 [pg/ml]	S	S	Stim3	Stim3	S	S	S	S	S	S

S sample [µM], *TA* Total activity, *NSB* Nonspecific binding, *Stim* maximal stimulation, *Basal* basal level
Typical quality control parameters:
  TA = 2.815 × 10 = 28.15
  %NSB = 0.02%
  %B0/TA = 2.25%
  Quality of fit = 1.00 (calculated from four parameter logistic curve fit)

3.3.10. Calculation of PGE₂
Concentration

(see Subheading 3.1.3.)

1. As Cremophor-EL/ethanol is of high-viscosity accordant volume must be pipetted fast. Keep the pipet horizontal.

2. All components for the PGE_2 ELISA are stable at 4°C until the expiration date; the conjugate and standard must be stored frozen at –20°C. The components have to be prepared as described by the company's manual and must be brought to room temperature for at least 30 min prior to opening.

3. All components for the LTB_4 ELISA are stable at 4°C until the expiration date; the conjugate must be stored frozen at –20°C. The components have to be prepared as described by the company's manual and must be brought to room temperature for at least 30 min prior to opening.

4. Wash the tip twice with blood before pipetting the blood into the caps. Pipet to the first arrest to avoid getting bubbles.

5. The standard and samples should be added into the wells fast. The samples also contain the compounds for maximal stimulation and basal data.

6. Every well should be green in colour except the NSB wells, which should be blue. The Blank and TA wells are empty at this point and have no color.

7. The different layers can be pipetted into the falcon tubes before taking the blood. X falcon tubes are necessary per day.

8. Pipet the second solution curtly under the polystyrene platelet.

9. Starting with 16 falcon tubes (4 falcons per donor), they are reduced after the first step to eight tubes (2 falcons per donor).

10. Sometimes platelets can be at the top of the solution. In this case, start removing the supernatant in the middle of the solution.

11. Platelet suspension of all donors is put in one falcon tube (15 ml).

12. The incubation time must not be longer than 5 min, because the trypanblue solution is cytotoxic. Otherwise, the count of dead cell is misrepresentative.

13. The tip should be placed at the coverslip's border; take care that the liquid does not leave the chamber.

14. The falcon tube should be turned after each step to get a homogeneous cell count in each tube.

15. Falcon tubes have to be mixed carefully to ensure a homogen cell count.

16. This solution has to be prepared fresh at the day of running out the assay.

17. All solutions must be prepared with bidistillated water.

18. The remaining blood for following samples must also be stored on the titramax at 37°C.

19. Wash the tip three times. Change tip after each step.

20. Take the accordant caps out of the water bath, wash the tip out in the blood suspension, mix the dilution carefully, and put the caps back into the water bath.

21. Diluted standards should be used within 60 min of preparation.

22. This stops the reaction and the plate should be read immediately.

23. Taking two different donors gives a better quality of results (because of reproducibility). Taking more than two donors might be too expensive.

24. Change tip after each step.

25. Use a multipipet to put the Acetylsalicylic acid into the caps.

26. Solution must be prepared fresh at the day of the experiment.

27. It is advisable to test also a reference compound.

28. Mix the samples intensively before use. Wash the tip out in the Assay buffer.

29. These parameters do not have to be measured each time.

30. Gloves should be worn.

31. Only eight compounds can be incubated at the same time.

32. There are 45 s time between starting the reaction with Calciumionophor dilution and stopping the reaction. That is why only three compounds can be incubated at the same time.

33. Allow the conjugate to warm to room temperature. Any unused conjugate should be aliquoted and refrozen at or below −20°C. Avoid repeated freeze–thaws of the aliquots.

34. Avoid repeated freeze–thaws.

35. Samples do not have to be diluted before use.

36. Taking four different donors gives a better quality of results (because of reproducibility). Taking more than two donors might be too expensive.

4. Notes

1. All solutions must be prepared with bidistilled water.

2. As Cremophor-EL/ethanol is of high-viscosity accordant volume must be pipetted fast. Keep the pipet horizontal.

3. Solution must be prepared fresh at the day of the experiment.

4. This solution has to be prepared fresh at the day of running out the assay.

5. All components for the PGE$_2$ ELISA are stable at 4°C until the expiration date; the conjugate and standard must be stored frozen at −20°C. The components have to be prepared as described by the company's manual and must be brought to room temperature for at least 30 min prior to opening.

6. Avoid repeated freeze–thaws.

7. All components for the LTB$_4$ ELISA are stable at 4°C until the expiration date; the conjugate must be stored frozen at −20°C. The components have to be prepared as described by the company's manual and must be brought to room temperature for at least 30 min prior to opening.

8. Allow the conjugate to warm to room temperature. Any unused conjugate should be aliquoted and refrozen at or below −20°C. Avoid repeated freeze–thaws of the aliquots.

9. Wash the tip twice with blood before pipetting the blood.

10. The remaining blood for following samples must also be stored on the titramax at 37°C.

11. Wash the tip three times. Change tip after each step.

12. Change tip after each step.

13. Take the accordant caps out of the water bath, wash the tip out in the blood suspension, mix the dilution carefully, and put the caps back into the water bath.

14. It is advisable to test also a reference compound.

15. Diluted standards should be used within 60 min of preparation.

16. Mix the samples intensively before use. Wash the tip out in the Assay buffer.

17. The standard and samples should be added into the wells fast. The samples also contain the compounds for maximal stimulation and basal data.

18. Every well should be green in colour except the NSB wells, which should be blue. The Blank and TA wells are empty at this point and have no color.

19. This stops the reaction and the plate should be read immediately.

20. These parameters don not have to be measured each time.

21. Use a multipipet to put the Acetylsalicylic acid into the caps.

22. The different layers can be pipetted into the falcon tubes before taking the blood.

23. Pipet the second solution curtly under the polystyrene platelet.

24. Starting with 16 falcon tubes (four falcons per donor), they are reduced after the first step to eight tubes (two falcons per donor).

25. Sometimes platelets can be at the top of the solution. In this case, start removing the supernatant in the middle of the solution.

26. Platelet suspension of all donors is put in one falcon tube (15 ml).

27. Gloves should be worn.

28. The incubation time must not be longer than 5 min, because the trypan blue solution is cytotoxic. Otherwise, the count of dead cell is misrepresentative.

29. The tip should be placed at the coverslip's border; take care that the liquid does not leave the chamber.

30. The falcon tube should be turned after each step to get a homogeneous cell count in each tube.

31. Only eight compounds can be incubated at the same time.

32. Falcon tubes have to be mixed carefully to ensure a homogen cell count.

33. There are 45 s between starting the reaction with Calciumionophor and stopping the reaction. This is why only three compounds can be incubated at the same time.

34. Samples do not have to be diluted before use.

References

1. Komhoff M, Grone HJ, Klein T, Seyberth HW, Nusing RM (1997) Localization of cyclooxygenase-1 and -2 in adult and fetal human kidney: implication for renal function. Am J Physiol 272:F460–F468

2. Funk CD, Funk LB, Kennedy ME, Pong AS, Fitzgerald GA (1991) Human platelet/erythroleukemia cell prostaglandin G/H synthase: cDNA cloning, expression, and gene chromosomal assignment. FASEB J 5:2304–2312

3. Goppelt-Struebe M (1995) Regulation of prostaglandin endoperoxide synthase (cyclooxygenase) isozyme expression. Prostaglandins Leukot Essent Fatty Acids 52:213–222

4. McGeer PL, McGeer E, Rogers J, Sibley J (1990) Anti-inflammatory drugs and Alzheimer's disease. Lancet 335:1037

5. Reines SA, Block GA, Morris JC, Liu G, Nessly ML, Lines CR, Norman BA, Baranak CC (2004) Rofecoxib Protocol 091 Study Group: Rofecoxib: no effect on Alzheimer's disease in a 1-year, randomized, blinded, controlled study. Neurology 62:66–71

6. Martyn C (2003) Anti-inflammatory drugs and Alzheimer's disease. Brit Med J 327:353–354

7. Patrono C, Ciabattoni G, Pinca E, Pugliese F, Castrucci G, De Salvo A, Satta MA, Peskar BA (1980) Low dose aspirin and inhibition of thromboxane B2 production in healthy subjects. Thromb Res 17:317–327

8. Laneuville O, Breuer DK, Dewitt DL, Hla T, Funk CD, Smith WL (1994) Differential inhibition of human prostaglandin endoperoxide H synthases-1 and -2 by nonsteroidal anti-inflammatory drugs. J Pharmacol Exp Ther 271:927–934

9. Riendeau D, Percival MD, Boyce S, Brideau C, Charleson S, Cromlish W, Ethier D, Evans J, Falgueyret JP, Ford-Hutchinson AW, Gordon R, Greig G, Gresser M, Guay J, Kargman S, Leger S, Mancini JA, O'Neill G, Ouellet M, Rodger IW, Therien M, Wang Z, Webb JK, Wong E, Xu L, Young RN, Zamboni R, Prasit P, Chan CC (1997) Biochemical and pharmacological profile of a tetrasubstituted furanone as a highly selective COX-2 inhibitor. Br J Pharmacol 121:105–117

10. Argentieri DC, Ritchie DM, Ferro MP, Kirchner T, Wachter MP, Anderson DW, Rosenthale ME, Capetola RJ (1994) Tepoxalin: a dual cyclooxygenase/5-lipoxygenase inhibitor of arachidonic acid metabolism with potent anti-inflammatory activity and a favorable gastrointestinal profile. J Pharmacol Exp Ther 271:1399–1408

11. Laufer S, Zechmeister P, Klein T (1999) Development of an in-vitro test system for the evaluation of cyclooxygenase-2 inhibitors. Inflamm Res 48:133–138

12. Patrignani P, Panara MR, Greco A, Fusco O, Natoli C, Iacobelli S, Cipollone F, Ganci A, Creminon C, Maclouf J, Patrono C (1994) Biochemical and pharmacological characterization of the cyclooxygenase activity of human blood prostaglandin endoperoxide synthases. J Pharmacol Exp Ther 271:1705–1712

13. Brideau C, Kargman S, Liu S, Dallob AL, Ehrich EW, Rodger IW, Chan CC (1996) A human whole blood assay for clinical evaluation of biochemical efficacy of cyclooxygenase inhibitors. Inflamm Res 45:68–74

14. Young JM, Panah S, Satchawatcharaphong C, Cheung PS (1996) Human whole blood assays for inhibition of prostaglandin G/H synthases-1 and -2 using A23187 and lipopolysaccharide stimulation of thromboxane B2 production. Inflamm Res 45:246–253

15. Prostaglandin E_2 Correlate-EIA™ Kit, assay designs, Ann Arbor

16. Leukotriene B_4 Correlate-EIA™ Kit, assay designs, Ann Arbor

Chapter 9

In Vitro Cyclooxygenase Activity Assay in Tissue Homogenates

Samir S. Ayoub

Abstract

The use of in vitro cyclooxygenase (COX) activity assays has been particularly useful for comparing the effect of different drugs on COX activity in different tissues. In addition, this assay is relatively quick, cheap, and a large number of samples can be tested at the same time. However, one limitation of this assay is the fact that it does not discriminate between the activities of different COX isoforms.

Key words: COX activity, Adrenaline, Glutathione, Arachidonic acid

1. Introduction

Cyclooxygenase (COX) isoenzymes are involved in a large number of physiological as well as pathological processes in almost all body organs and fluids (1). Therefore, it is important to be able to measure the activity of these enzymes in different tissues. Determination of prostanoid concentration is used as a measure of COX activity. COX activity can be measured in tissues after extraction of prostanoids with C18 columns (2, 3). This protocol is often used to examine COX activity in different pathophysiological conditions and also to test the effect of drugs on in vivo synthesis of prostanoids in different tissues (4, 5). However, comparing the effect of drugs on COX activity in different tissues in vivo is complicated by pharmacokinetic factors. Therefore, the use of in vitro COX activity assays, described in this chapter, for comparison of the relative potencies of drugs on the inhibition of COX activity in different tissues is a more appropriate approach (6).

Samir S. Ayoub et al. (eds.), *Cyclooxygenases: Methods and Protocols*, Methods in Molecular Biology, vol. 644, DOI 10.1007/978-1-59745-364-6_9, © Springer Science+Business Media, LLC 2010

2. Materials

1. Homogenate buffer/Protease inhibitors: 50 mM Tris–HCl pH 7.4, 1 mM PMSF, 1.5 mM pepstatin A, and 0.2 mM leupeptin.
2. 50 mM Tris–HCl buffer pH 7.4.
3. Ultra-Turrax T25 homogenizer (Kanke and Kunkel IKA Labortechnik).
4. Bradford Reagent (Bio Rad).
5. Bovine serum albumin.
6. Tecan GENios plate reader (Jencans).
7. 50 mM Glutathione, reduced form.
8. Adrenaline.
9. 1.5 mM Arachidonic acid.

3. Methods

3.1. Tissue Collection

1. C57BL/6 mice are killed by cervical dislocation.
2. The whole spinal cord is ejected from the spinal column by initially making incisions on the dorsal and cranial ends of the spinal column then inserting a 23G needle attached to a saline-filled syringe at the dorsal end and forcing the whole spinal cord out. The spinal cord is immediately snap-frozen in liquid nitrogen.
3. The whole brain is removed from the skull and dissected anatomically into the cerebral cortex, mid brain, brain stem, and cerebellum and snap-frozen in liquid nitrogen (see Note 1).

3.2. Preparation of Tissues for COX Activity

3.2.1. Tissue Homogenization

1. Frozen samples are transferred to a tube containing 3 volumes of ice-cold homogenate buffer.
2. The samples are homogenized for 1 min with the tube placed in a beaker containing ice to keep the samples cool (see Notes 2 and 3).

3.2.2. Determination of Protein Concentration

1. The samples are spun down $1,000 \times g$ for 10 min at 4°C.
2. The supernatant is diluted in 50 mM Tris–HCl pH 7.4 at dilutions between 1:10 and 1:40.
3. 200 µl of Bradford reagent, diluted 1:5 in distilled water, is added to 10 µl sample in a 96-well plate.
4. The optical density is measured spectrophotometrically at 595 nm in a plate reader and protein concentrations determined from a bovine serum albumin standard curve read on the same plate at concentrations of 0.05, 0.1, 0.2, 0.3, 0.4, and 0.5 mg/ml.

3.3. COX Activity Assay

1. The COX activity assay is carried out in a total volume of 500 µl.

2. To all the reaction tubes 340 µl Tris–HCl (pH 7.4) is added (see Notes 4 and 5).

3. In addition, 50 µl of 50 mM reduced glutathione (5 mM final concentration) and 50 µl of 50 mM adrenaline (5 mM final concentration) are added (see Note 6).

4. 50 µl of sample is then added to all the tubes.

5. To start the reaction 10 µl of 1.5 mM arachidonic acid (30 µM final concentration) is added.

6. Immediately all tubes are wrapped with aluminium foil and placed in a water bath set at 37°C for 30 min.

7. The reaction is stopped after 30 min by boiling the samples at 95°C for 10 min.

8. The samples are then spun down at 3,000×g for 10 min at 4°C.

9. Finally the samples are stored at –80°C ready for prostanoid measurement.

4. Notes

1. This assay can also be used to measure COX activity in other organs, cultured cells and in body fluids such as inflammatory exudates.

2. Samples need to be kept on ice during the entire procedure to prevent protein degradation.

3. The head of the homogenizer needs to be washed thoroughly with distilled water between samples.

4. If different samples are used as source of COX activity, the amount of sample added to the reaction mixture is normalized to amount of protein, hence the volume of sample added would be different between different samples and the volume of Tris–HCl buffer is altered accordingly. 20 µg of protein is ideal.

5. If the assay is to be used to test the inhibitory effect of drugs on COX activity, 50 µl of drug, at the appropriate concentrations, is added to the reaction mixture and the volume of Tris–HCl is altered accordingly to give a final volume of 500 µl.

6. Reduced glutathione and adrenaline are important co-factors for COX activity.

References

1. Vane JR, Bakhle YS, Botting RM (1998) Cyclooxygenases 1 and 2. Annu Rev Pharmacol Toxicol 38:97–120
2. Ayoub SS, Botting RM, Goorha S, Colville-Nash PR, Willoughby DA, Ballou LR (2004) Acetaminophen-induced hypothermia in mice is mediated by a prostaglandin endoperoxide synthase 1 gene-derived protein. Proc Natl Acad Sci U S A 101(30): 11165–11169
3. Ayoub SS, Colville-Nash PR, Willoughby DA, Botting RM (2006) The involvement of a cyclooxygenase 1 gene-derived protein in the antinociceptive action of paracetamol in mice. Eur J Pharmacol 538(1–3):57–65
4. Hetu PO, Riendeau D (2005) Cyclooxygenase-2 contributes to constitutive prostanoid production in rat kidney and brain. Biochem J 391(Pt. 3):561–566
5. Yoshikawa K, Kita Y, Kishimoto K, Shimizu T (2006) Profiling of eicosanoid production in the rat hippocampus during kainic acid-induced seizure: dual phase regulation and differential involvement of COX-1 and COX-2. J Biol Chem 281(21):14663–14669
6. Mitchell JA, Kohlhaas KL, Sorrentino R, Warner TD, Murad F, Vane JR (1993) Induction by endotoxin of nitric oxide synthase in the rat mesentery: lack of effect on action of vasoconstrictors. Br J Pharmacol 109(1):265–270

Chapter 10

Techniques Used to Characterize the Binding of Cyclooxygenase Inhibitors to the Cyclooxygenase Active Site

William F. Hood

Abstract

Inhibitors of enzyme-catalyzed reactions are typically characterized by their ability to diminish product formation while altering the Michaelis Menten constants V_{max} and K_m. Determination of an apparent inhibitor affinity (K_i) for the enzyme is also possible using this approach. Unfortunately, analysis of product formation does not easily provide information regarding the kinetics of inhibitor binding and may not be possible depending upon the mechanism of action. Radiolabeling of the inhibitor allows one to do a direct binding assay and thereby more directly determine the kinetics of inhibitor binding. With this in mind, we developed a radioligand-based binding assay for inhibitors of cyclooxygenase.

Key words: Cyclooxygenase, Binding, Kinetics, Radiolabeled, Inhibitor

1. Introduction

Cyclooxygenase (COX) is a heme-containing enzyme that catalyzes the conversion of arachidonic acid into prostaglandin H_2 (1, 2). There appear to be two COX isozymes: a constitutive form, COX-1, and an inducible form, COX-2, which produce prostaglandins for different functions. COX-1 prostaglandins are important in gastric protection and platelet aggregation while COX-2 prostaglandins are primarily involved in pain management and inflammation. Early inhibitors of cyclooxygenase, such as the nonsteroidal anti-inflammatory drugs (NSAIDS), were developed but they produced some gastrointestinal (GI) toxicity. The GI disruption primarily was a result of the NSAID inhibition of COX-1. With the hopes of reducing unwanted GI side effects while maintaining

Samir S. Ayoub et al. (eds.), *Cyclooxygenases: Methods and Protocols*, Methods in Molecular Biology, vol. 644,
DOI 10.1007/978-1-59745-364-6_10, © Springer Science+Business Media, LLC 2010

the anti-inflammatory properties, development of NSAIDS having a greater selectivity for COX-2 inhibition became paramount.

The early marketed NSAIDS were found to inhibit COX-1 and COX-2 with varying potencies and selectivity (3, 4). However, the selectivity and potency for COX-1 and COX-2 as measured by enzyme activity and product formation varied wildly depending upon the assay system (5–7). Notwithstanding, highly selective and potent inhibitors of COX-2, were discovered. To potentially better understand how the compounds demonstrated their COX-2 selectivity, several investigators focused on investigating the kinetics of the inhibitor binding. Lanzo et al. (8) used a fluorescent diaryl heterocyclic oxazole inhibitor and followed the quench upon binding to COX. Others (9, 10) have used a mathematical model to discern the kinetic rate constants of inhibitor binding. We chose to radiolabel several COX-2 selective inhibitors, including celecoxib and valdecoxib, and developed a radioligand binding assay using these compounds to more directly characterize their interaction (especially kinetics) with the cyclooxygenase active site (11) and provide a hypothesis for their selectivity.

2. Materials

2.1. Immobilizing Cyclooxygenase

1. 96-well Immulon 2 microtiter plates (Dynex Technologies Inc., Chantilly, VA) (see Note 1).

2. Antibody dilution buffer: 100 mM $NaHCO_3$, pH 8.2.

3. Plate washing solution: Dulbecco's phosphate buffer saline without $CaCl_2$ and $MgCl_2$, pH 7.4 (D-PBS; Invitrogen, Carlsbad, CA).

4. Blocking reagent: 10% skim milk in D-PBS; make fresh daily as needed.

5. COX dilution buffer: 100 mM Tris-HCl, 1 μM hemin (Sigma Chemical Co., St. Louis, MO), pH 8.0. Stock 1 mM hemin in dimethylsulfoxide (DMSO) can be stored as 0–4°C.

6. Murine monoclonal COX-1-specific antibody (M584-7f4) and COX-2-specific antibody (R6) were prepared in-house by PGRD.

7. COX-1 and COX-2 were expressed and purified as described earlier (12) (see Note 2).

2.2. Radioligand Incubation

1. Assay solution buffer: 100 mM Tris-HCl, 1 μM hemin, pH 8.0, with 5% DMSO (Sigma Chemical Co., St. Louis, MO).

2. Radiolabeled ligand: [^{3}H]-celecoxib (specific radioactivity of 3 Ci/mmol) or [^{3}H]-valdecoxib (10 Ci/mmol) (American Radiolabeled Chemicals Inc., St. Louis, MO).

3. Wash buffer: D-PBS, ice cold; store at 0–4°C.

4. Bound removal solution: 10% sodium dodecyl sulfate (SDS) in D-PBS.

3. Methods

The actual methodology for the binding assay is not particularly involved or tedious. In fact, the general procedure is applicable for most enzymes or receptors, assuming one has a good radioligand and antibody. A good radioligand would be one that has high affinity (<20 nM) and low background or nonspecific binding. A good antibody would be non-neutralizing and have sufficient avidity so that the captured enzyme does not readily dissociate. The difficulty in the assay resides more in performing the proper control experiments, data manipulation, and assay validation.

3.1. Immobilizing Cyclooxygenase

1. COX antibody is diluted in 100 mM NaHCO$_3$, pH 8.2 to a concentration of 10 μg/ml and 100 μl added to each microtiter plate well. Plates are incubated with antibody overnight at 25°C (see Note 3) in a humidified chamber (see Note 4).

2. The next morning, the antibody solution is removed from the plate by inverting the plate and shaking it. Fresh, room temperature D-PBS (~200 μl) is added to each well, the plate is inverted, and the D-PBS is shaken out. This plate-washing procedure is repeated three to four times, leaving a plate with immobilized antibody.

3. To the antibody coated plate is added 200 μl of a solution of 10% skim milk in D-PBS. The coated plate with skim milk is incubated at 37°C for 90–120 min. After this period of time (longer or shorter time may work as well), the plate is washed with D-PBS as described above (see Note 5) leaving an antibody coated plate.

4. Next, 50 μl COX enzyme (10–50 μg/ml), diluted in assay buffer (100 mM Tris-HCl, 1 μM hemin, pH 8.0) is added to the antibody coated wells (see Note 6) and incubated at 25°C for 60–120 min.

5. Finally, the antibody and COX coated wells are washed with D-PBS as described above. Any residual liquid should be gently aspirated off so there is minimal dilution to the binding solution upon its addition in the next step.

3.2. Radioligand Incubation

1. To the antibody-COX coated plates is added 85 μl of assay binding solution (100 mM Tris-HCl, 1 μM hemin, pH 8.0), 5 μl DMSO (with or without unlabeled test compound), and 10 μl of [³H]-ligand. The radiolabeled ligand is added last to initiate the incubation binding assay (see Note 7). When all components have been added, the plate should be put on a shaker to continuously mix the incubation.

2. To halt the binding incubation, each individual well is rapidly aspirated and washed twice (<2 s) with 250 μl ice cold D-PBS, leaving only the COX bound radioligand in the wells.

3. To remove the COX bound radioligand from the well, 50 μl of 10% SDS in D-PBS is added. The plates are then allowed to sit at room temperature for at least 30 min before this 50 μl SDS solution is removed by pipet and transferred into a 7 ml liquid scintillation vial (see Note 8) for quantitation using liquid scintillation spectrometry.

3.3. Coated Plate Assay Validation

Validation consists of a series of experiments designed to ensure that the assay provides a true indication of inhibitor binding to a relevant pharmacological site.

1. The best initial experiment is a dissociation time course. Incubate a low concentration (1–10 nM) of radioligand with COX for 2–4 h at 25°C. After this time, aspirate off the liquid containing the unbound, free [³H]-ligand, leaving behind in the plate the COX bound [³H]-ligand. To the plate, add 150 μl of excess cold (~100× K_D; e.g. 10 μM), unlabeled compound. Addition of the excess unlabeled compound initiates the dissociation time course. At various times after the addition of this excess compound, aspirate the well. Do not add a wash solution at this time during this initial dissociation time course. Remove the bound counts as described above using 10% SDS (see Note 9).

2. Association time course – For association time course experiments, after the addition of [³H]-labeled ligand, the incubations are halted at various time points by aspiration and rapid washing (see Note 10).

3. Saturation experiment – Incubate various concentrations of [³H]-ligand with the immobilized COX. Then, after a specified period of incubation time (see Note 11), aspirate and rapidly wash the wells. Ideally, the concentrations should span the expected K_D; for example, from 0.1 to 10 times the K_D (Fig. 1).

4. Unlabeled compound inhibition assays – For the competitive binding assays, a second 96 well plate is used to initially make the dose response dilutions of the compound (see Note 12). To each of these different concentrations, a consistent trace

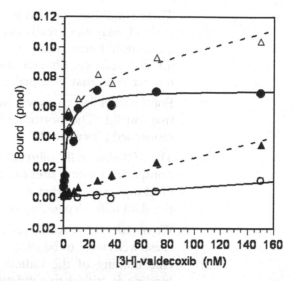

Fig. 1. Saturation binding curve of [³H]-valdecoxib to either COX-1 (*open circles*) or COX-2 (*filled circles*). Plates were coated with either COX-1 or COX-2 and then incubated with various concentrations of [³H]-valdecoxib for 240 min at room temperature. The free unbound radioligand was aspirated off, the plate rapidly washed and the remaining amount bound determined using liquid scintillation spectrometry. Data shown are from one representative experiment using [³H]-valdecoxib. For clarity purposes, only the non-specific (*filled triangles*) and total (*open triangles*) binding to COX-2 are shown.

amount of [³H]-labeled ligand is added and then the complete mixture of cold test compound and trace radiolabel is added to the enzyme coated plate. This method ensures that there is no preincubation of either the test compound or the radiolabeled ligand with the enzyme (see Note 13).

4. Notes

1. The Immulon 2 plates worked well for these particular radioligands but other microtiter plates may work just as well or even better for other radiolabeled compounds. We found significant differences in the amount of nonspecific binding between different radiolabeled inhibitors and different microtiter plates.

2. To ensure COX has been immobilized and remains immobilized on the plate for the length of the binding assay, one should perform an enzyme activity check by following the production of PGE2 upon 10 μM arachidonic acid (Nu-Chek Prep, Inc., Elysian, MN) addition to the plate. Several companies supply PGE2 ELISA kits. We used in-house reagents.

3. These conditions worked for our assay. The concentration of antibody may vary greatly depending upon antibody and plate interaction. Furthermore, overnight at room temperature was for convenience. Shorter incubation periods and different temperatures may be equally effective.

4. For a humidified chamber, we used a plastic container with a snap-on lid. The bottom of the container was lined with moistened paper towels.

5. The 10% skim milk solution is added to help block potential nonspecific binding sites on the plate. There are various other blocking solutions that may work as well or even better than the skim milk, depending upon your system.

6. We opted to leave some wells without the COX addition since these wells were to be used as a measurement of the nonspecific binding of the radioligand. Definition of nonspecific binding is critical to validating the assay. Although radiolabeled celecoxib and valdecoxib possessed little or no high affinity binding for the plate or antibody, we did have issues with other radiolabeled compounds that we examined. Other possibilities that could help define the nonspecific binding component would be wells that have either (a) no antibody, no COX, (b) antibody, no COX or (c) antibody, COX, excess cold (~100× K_D) unlabeled ligand.

7. These binding studies were performed at 25°C. Lower (e.g., 0–4°C) or higher (e.g., 37°C) temperatures may work better depending upon the enzyme and radioligand.

8. We chose to remove the entire 50 µl SDS solution and use individual 7 ml liquid scintillation vials for counting purposes. One could easily use multichannel pipets or robotics to remove all or a portion of the SDS and transfer into a scintillant coated microtiter plate for higher throughput of the assay.

9. This initial dissociation time course experiment will allow one to determine the time ($t\,\frac{1}{2}$) necessary for half of the radioactivity to dissociate from the COX. Determining this time at the outset then gives one confidence that when subsequent samples are rapidly washed with cold D-PBS there is little or no loss of specific binding during the time taken for the wash. A more extensive time course, including washes of the samples, will then provide the dissociation rate constant (k_{off} or k_{-1}). Data can be best fit to the equation $(RL) = (RL)_0\, e^{(-k \times t)}$ where t is time and k is the dissociation rate constant (Fig. 2).

10. For best results when performing the association time course, use multiple concentrations (e.g., three to six) of the radioligand and use a software program providing global analysis of

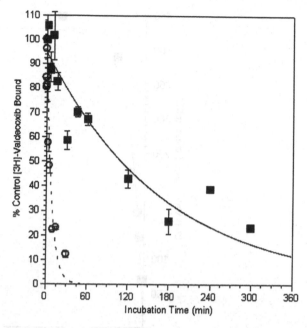

Fig. 2. Dissociation time course for [³H]-valdecoxib binding to COX-2 (*filled squares*) and a COX mutant, V523I (*open circles*). Plates were coated with either wild type COX-2 or the mutant V523I and incubated with [³H]-valdecoxib for 120 min. Following this initial incubation period, the free unbound radioligand was aspirated off and 150 μl of excess cold ligand (10 μM) added to initiate the dissociation time course. Then, at various times the wells were aspirated and the remaining amount bound determined using liquid scintillation spectrometry. Data shown are the mean ± SEM from at least three experiments each performed in triplicate.

the data. This type of experiment will yield the association rate constant (k_{on} or k_{+1}) (Fig. 3.).

11. From the association time course experiment one should know how long to incubate the samples until they reach equilibrium, or steady state. Obtaining equilibrium is necessary for proper calculation of the saturation K_D. Data can be best fit to the equation $(RL) = (R)_t (L)/(K_D + L)$ where (RL) is the amount bound, $(R)_t$ is the total enzyme concentration, L is the ligand concentration, and K_D is the equilibrium dissociation constant. This equation can only be used if free, unbound ligand approximates the total ligand added. The resulting saturation K_D should be similar to the kinetically derived K_D as determined from k_{off}/k_{on} (see Table 1).

12. For assay validation the first compounds tested should be those which are standard in other associated assays. For COX, we chose various NSAIDs. The rank order potency in the binding should mimic that in other assays.

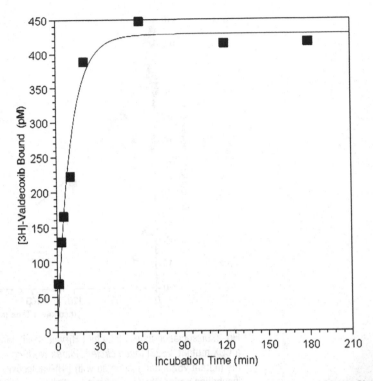

Fig. 3. Association time course for [³H]-valdecoxib binding to wild type COX-2. Plates coated with wild type COX-2 were incubated with [³H]-valdecoxib (5 nM) for various lengths of time. Following the incubation period, the free unbound radioligand was aspirated, the plate rapidly washed and the remaining amount bound determined using liquid scintillation spectrometry. Data shown are from one representative experiment; $k+1 = 4.0 \times 10^6\,M^{-1}\,min^{-1}$.

Table 1
Equilibrium and kinetic constants for [³H]-valdecoxib binding to COX-2

Saturation K_D	3.2 ± 0.9 nM (5)
Dissociation rate ($\times 10^{-3}\,min^{-1}$)	6.9 ± 0.1 (5)
Association rate ($\times 10^6\,min^{-1}\,M^{-1}$)	4.5 ± 0.6 (3)
Kinetic K_D	1.5 nM

Data are the mean ± SEM of the number of experiments shown in parentheses from experiments which were performed in triplicate

13. K_i values for inhibitors may be derived from the concentration of competing compound that inhibits 50% of the binding observed in the absence of any inhibitor (IC_{50}) using the equation of Cheng and Prusoff (13).

References

1. Smith WL, Garavito RM, DeWitt DL (1996) Prostaglandin endoperoxide H synthases (cyclooxygenases)-1 and -2. J Biol Chem 271(52):33157–33160

2. Marnett LJ, Rowlinson SW, Goodwin DC, Kalgutkar AS, Lanzo CA (1999) Arachidonic acid oxygenation by COX-1 and COX-2. Mechanisms of catalysis and inhibition. J Biol Chem 274(33):22903–22906

3. Patrignani P, Panara MR, Sciulli MG, Santini G, Renda G, Patrono C (1997) Differential inhibition of human prostaglandin endoperoxide synthase-1 and -2 by nonsteroidal anti-inflammatory drugs. J Physiol Pharmacol 48(4):623–631

4. Warner TD, Giuliano F, Vojnovic I, Bukasa A, Mitchell JA, Vane JR (1999) Nonsteroid drug selectivities for cyclo-oxygenase-1 rather than cyclo-oxygenase-2 are associated with human gastrointestinal toxicity: a full in vitro analysis. Proc Natl Acad Sci USA 96(13):7563–7568

5. Meade EA, Smith WL, DeWitt DL (1993) Differential inhibition of prostaglandin endoperoxide synthase (cyclooxygenase) isozymes by aspirin and other non-steroidal anti-inflammatory drugs. J Biol Chem 268(9):6610–6614

6. Mitchell JA, Akarasereenont P, Thiemermann C, Flower RJ, Vane JR (1993) Selectivity of nonsteroidal antiinflammatory drugs as inhibitors of constitutive and inducible cyclooxygenase. Proc Natl Acad Sci USA 90(24):11693–11697

7. Gierse JK, Koboldt CM, Walker MC, Seibert K, Isakson PC (1999) Kinetic basis for selective inhibition of cyclo-oxygenases. Biochem J 339(Pt 3):607–614

8. Lanzo CA, Sutin J, Rowlinson S, Talley J, Marnett LJ (2000) Fluorescence quenching analysis of the association and dissociation of a diarylheterocycle to cyclooxygenase-1 and cyclooxygenase-2: dynamic basis of cyclo-oxygenase-2 selectivity. Biochemistry 39(20): 6228–6234

9. Callan OH, So OY, Swinney DC (1996) The kinetic factors that determine the affinity and selectivity for slow binding inhibition of human prostaglandin H synthase 1 and 2 by indomethacin and flurbiprofen. J Biol Chem 271(7):3548–3554

10. So OY, Scarafia LE, Mak AY, Callan OH, Swinney DC (1998) The dynamics of prostaglandin H synthases. Studies with prostaglandin H synthase 2 Y355F unmask mechanisms of time-dependent inhibition and allosteric activation. J Biol Chem 273(10):5801–5807

11. Hood WF, Gierse JK, Isakson PC, Kiefer JR, Kurumbail RG, Seibert K, Monahan JB (2003) Characterization of celecoxib and valdecoxib binding to cyclooxygenase. Mol Pharmacol 63:870–877

12. Gierse JK, Hauser SD, Creely DP, Koboldt C, Rangwala SH, Isakson PC, Seibert K (1995) Expression and selective inhibition of the constitutive and inducible forms of human cyclooxygenase. Biochem J 305:479–484

13. Cheng Y, Prusoff WH (1973) Relationship between the inhibition constant (K_i) and the concentration of inhibitor which causes 50 per cent inhibition (IC_{50}) of an enzymatic reaction. Biochem Pharmacol 22: 3099–3108

Chapter 11

In Vitro Assays for Cyclooxygenase Activity and Inhibitor Characterization

Mark C. Walker and James K. Gierse

Abstract

Cyclooxygenases (COX), or Prostaglandin H Synthases (PGHS), are the target enzymes for nonsteroidal anti-inflammatory drugs (NSAIDS). The identification of two isoforms of COX nearly 20 years ago stimulated a flurry of research activity to identify novel, selective inhibitors that could provide potential benefit over existing nonselective NSAIDS. An important contribution to this discovery effort was the development of various *in vitro* and *in vivo* assays to support rapid screening of chemical libraries, characterization of inhibitory mechanism, and determination of potency and efficacy to guide follow-up medicinal chemistry efforts. Several assay methods for the *in vitro* evaluation of COX activity and mechanism of inhibition by test compounds will be reviewed. Each of these methods has inherent advantages and disadvantages with regard to application and the mechanistic detail provided.

Key words: Cyclooxygenase, COX, Prostaglandin H synthase, PGH synthase, PGHS, Prostaglandin, PGE, PGG, PGH, Nonsteroidal anti-inflammatory drugs, NSAIDs, Oxygen uptake, Oxygen electrode, Oxygraph, Peroxidase, ELISA

1. Introduction

Cyclooxygenases (COX), or more specifically Prostaglandin H_2 Synthases (PGHS), catalyze the first committed step in the biosynthesis of the prostaglandins (PGs) and thromboxanes (Txs) and are the pharmacological target of nonsteroidal anti-inflammatory drugs (NSAIDs) (see (1–5) for review). Two isoforms of COX have been identified whose expression is subject to distinct transcriptional regulation. One isoform, designated COX-1, is expressed in nearly all tissues and is thought to be responsible for the production of prostanoids at basal levels for the maintenance of physiological homeostasis (6). Another isoform, designated

Samir S. Ayoub et al. (eds.), *Cyclooxygenases: Methods and Protocols*, Methods in Molecular Biology, vol. 644,
DOI 10.1007/978-1-59745-364-6_11, © Springer Science+Business Media, LLC 2010

COX-2, is expressed in response to inflammatory stimuli and is primarily responsible for the dramatic increase in PG production observed at sites of inflammation. The inhibition of COX-2-derived PG production is thought to be responsible for the antipyretic, anti-inflammatory, and analgesic properties of NSAIDs (7, 8). Adverse side effects, such as increased incidence of gastric ulceration and renal complications, associated with the chronic administration of NSAIDs are thought to arise as a consequence of the inhibition of COX-1 (9). Most NSAIDs are relatively nonselective, inhibiting COX-1 as well as COX-2 (10, 11). The discovery of distinct COX isoforms and the subsequent cloning and expression of these isoforms has contributed to the development of new inhibitors that are selective for COX-2 (12–16).

The COX enzymes catalyze the bis-oxygenation of arachidonic acid to produce PGH_2 and are, therefore, more appropriately referred to as Prostaglandin H_2 Synthases (PGHS). The catalytic cycle proceeds via two distinct steps (Fig. 1), which occur at distinct active sites (17). The first step, referred to as the cyclooxygenase reaction, is initiated by the abstraction of a hydrogen atom from arachidonic acid by an enzyme-based radical, resulting in the incorporation of two moles of molecular oxygen to yield the hydroperoxide, PGG_2, as the initial product. The second step, referred to as the peroxidase reaction, catalyzes the reduction of the hydroperoxide of PGG_2 to form the corresponding alcohol, PGH_2. The mechanism for the peroxidase reaction is typical of the heme-containing peroxidases, in which the reduction of a peroxide substrate is coupled to the oxidation of a small molecule co-substrate.

To support the discovery and development of novel COX-2 selective drugs, a variety of *in vitro* assays have been developed for the rapid screening and characterization of new leads. Each of these assay methods has advantages and disadvantages with respect to ease of automation and screening of large chemical libraries, evaluation of protein expression or site-directed mutations, and providing mechanistic detail (Fig. 2).

The most direct method for quantitating COX activity is the oxygraph assay, which utilizes a Clark-style oxygen electrode to continuously monitor the consumption of molecular oxygen during enzymatic turnover at the cyclooxygenase active site. The oxygraph assay provides a determination of initial rate and is appropriate for the analysis of both steady state and time-dependent inhibitory mechanisms. However, the method is timeconsuming and, therefore, is very low throughput, which makes it unsuitable for the rapid screening of chemical libraries. Furthermore, the oxygraph assay suffers from a lack of sensitivity, requiring relatively high concentrations of enzyme and inhibitor.

As an alternative to the oxygraph assay, the peroxidase assay measures the formation of PGH_2 by coupling the reduction of the initial peroxide product, PGG_2, to the oxidation of an optically

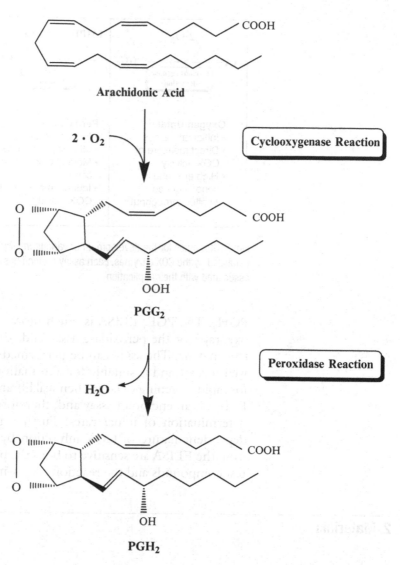

Arachidonic Acid

$2 \cdot O_2$

Cyclooxygenase Reaction

PGG$_2$

H_2O

Peroxidase Reaction

PGH$_2$

Fig. 1. Two distinct steps along the Prostaglandin H$_2$ Synthase reaction pathway. The first step is referred to as the cyclooxygenase reaction, which catalyzes the incorporation of two moles of molecular oxygen into arachidonic acid to yield the bis-endohydroperoxide intermediate, PGG$_2$. The second step is referred to as the peroxidase step, which catalyzes the two-electron reduction of the hydroperoxide moiety of PGG$_2$ to the alcohol to yield the final product, PGH$_2$.

active redox dye, such as N,N,N',N'-tetramethyl-p-phenylenediamine (TMPD), as co-substrate. The peroxidase assay can be utilized either as an endpoint assay or as a continuous assay for the determination of initial rates, and is readily adapted to a 96-well plate format. The peroxidase assay is more sensitive than the oxygraph assay, but it is an indirect measure of turnover at the cyclooygenase-active site.

As a third alternative, the enzymatic conversion of arachidonic acid to PGH$_2$ can be conveniently measured by ELISA detection of PGE$_2$, the non-enzymatic decomposition product of

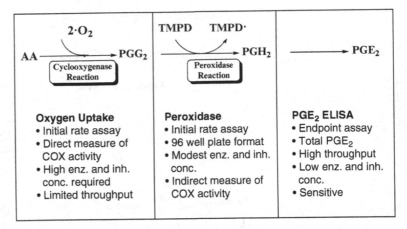

Fig. 2. Strategies for the development of various *in vitro* assays to monitor the reaction catalyzed by the COX enzymes. Each assay method has advantages and disadvantages associated with their application.

PGH$_2$. The PGE$_2$ ELISA is much more sensitive than either the oxygraph or the peroxidase assay and, therefore, requires much less enzyme. The assay can be performed in either a 96- or 384-well format and is suitable for automation, which makes it ideal for rapid screening of large chemical libraries. However, the PGE$_2$ ELISA is an end point assay and, therefore, is not suitable for the determination of initial rates. Furthermore, due to the time-dependent nature of COX inhibition by most NSAIDs, results from the ELISA are sensitive to both the pre-incubation time with test compounds and the reaction time with substrate.

2. Materials

2.1. Enzymatic Assays

Unless stated otherwise, all reagents are obtained from Sigma–Aldrich, St. Louis, MO. Water is milli-Q purified.

1. COX-1 from sheep seminal vesicles, and recombinant murine COX-2, were purified as described previously (18).

2. A 1.0 M stock solution of Tris buffer, pH 8.1 at 25°C, is prepared by mixing equal volumes of 1.0 M Tris-HCl with 1.0 M Tris base. The stock is filtered through a 0.2 μm CA membrane and stored at 5°C.

3. A 1.0 M stock solution of potassium phosphate buffer, pH 7.5, is prepared by dissolving 31.3 g monobasic potassium phosphate and 134.1 g dibasic potassium phosphate into 1 L water. The stock is filtered through a 0.2 μm CA membrane and stored at 5°C.

4. Oxygraph Assay buffer containing 0.1 M Tris, pH 8.1, 1 μM hemin, and approximately 0.5 mM phenol is prepared by

diluting 20 ml 1.0 M Tris, pH 8.1, into 180 ml air-saturated H_2O (see Note 1), to which is added 0.2 ml 1 mM hemin in DMSO and 1 drop (approx. 0.1 ml) of water-saturated phenol.

5. Peroxidase Assay buffer is prepared by diluting 2.2 ml 1.0 M Tris, pH 8.1, and 22 μl 1 mM hemin into 15.4 ml air-saturated H_2O (17.6 ml final). The final volume is sufficient for a single 96-well microtiter assay plate.

6. PGE_2 ELISA Assay buffer containing 50 mM potassium phosphate, pH 7.5, 2 μM phenol, and 1 μM hemin is prepared by diluting 5 ml 1.0 M potassium phosphate stock, pH 7.5, 10 μl 10 mM hemin in DMSO, and 10 μl phenol to a final volume of 100 ml with water.

7. Arachidonic acid (supplied as the sodium salt; Nu-Chek Prep, Elysian, MN) stock solutions of 10 mM are prepared by dissolving solid into water at 4 °C with the aid of a rocker plate. After the solid is fully dissolved, the stock solution is aliquoted as 0.5–1 ml volumes and stored at –80°C (see Note 2).

8. Water-saturated phenol is prepared by dissolving solid phenol (ultra-pure grade, Cat# 5509UA, BRL, Bethesda, MD) into excess water. The stock solution is stored in the dark at 4°C.

9. Stock solutions of 10 mM and 1 mM hemin are prepared in DMSO and aliquoted as 0.5 ml volumes for storage at –80°C.

10. N,N,N',N'-Tetramethyl-p-phenylenediamine dihydrochloride (TMPD) is prepared as a 17 mM stock in water and stored as 0.5 ml aliquots at –20°C (see Note 3).

11. Peroxidase Substrate stock is prepared by diluting 0.22 ml 17 mM TMPD and 0.22 ml 10 mM arachidonate into 1.76 ml H_2O (2.2 ml final). The final volume of this substrate-containing stock is sufficient for a single 96-well microtiter assay plate.

12. A 10 mM stock solution of indomethacin (Cat # 17378; Sigma–Aldrich, St. Louis, MO) is prepared in DMSO. The stock solution is stored at room temperature for no more than two or three weeks (see Note 4).

13. 96-well flat bottom microtiter plates (VWR Internatl., St. Louis, MO).

2.2. PGE₂ ELISA Reagents

1. EIA buffer is prepared by diluting 200 ml 1.0 M potassium phosphate stock, pH 7.5, into approximately 1.5 L water, with the addition of 200 mg sodium azide, 46.8 g sodium chloride, 740 mg tetrasodium EDTA, and 2 g bovine serum albumin (BSA; Cat. # A7030; Sigma–Aldrich, St. Louis, MO). After dissolution, the final volume is adjusted to 2.0 L with water.

2. Saturation buffer is prepared by the addition of 100 mg sodium azide and 1 g BSA to 500 ml EIA buffer. This provides sufficient volume for blocking fifty 96-well plates.

3. Plate washing buffer is prepared by diluting 100 ml 1.0 M potassium phosphate buffer and 5 ml Tween-20 to a final volume of 10 L with water, to which 1 g sodium azide is then added.

4. Mouse anti-PGE_2 monoclonal primary antibody (19) (Cat # 414013; Cayman Chemical, Ann Arbor, MI) is reconstituted as described by the supplier.

5. Nunc Immunoplates (Cat #439454; Nalge Nunc Internatl., Rochester, NY) are coated with affinity purified Goat antimouse IgG secondary antibody (Cat # 115-005-003; Jackson Immuno-Research Laboratories, Inc., West Grove, PA) at 8 µg in 200 µl per well in 50 mM potassium phosphate, pH 7.4, overnight at room temperature. The plates are then blocked from any further protein binding by addition of 100 µl saturation buffer per well. Saturation is usually complete after 6 h at room temperature. Plates are then wrapped in plastic wrap and stored at 4°C.

6. A stock solution of PGE_2 (Cat. # 14010; Cayman Chemical, Ann Arbor, MI) is prepared by dissolving 5.0 mg PGE_2 into 1 ml ethanol and storing at –80°C as 100 µl aliquots. Aliquots are thawed as needed and further diluted to 500 ng/ml with milli-Q water. This working stock is stored at –20°C and is used to construct the PGE_2 standard curve for the ELISA assay.

7. A stock solution of PGE_2-AChE tracer is prepared by dissolving the contents of 1 vial of PGE_2-Acetylcholinesterase (Cat. # 414010; Cayman Chemical, Ann Arbor, MI) into 100 ml EIA buffer.

8. Ellman's reagent (Cat #400050; Cayman Chemical, Ann Arbor, MI) is prepared immediately prior to use by dissolving 1 vial into 50 ml water.

9. 96-well round bottom microtiter plates (VWR Internatl., St. Louis, MO).

10. 96-well plate sealers (VWR Internatl., St. Louis, MO).

3. Methods

3.1. Oxygraph Assay

Enzymatic activity is monitored continuously by following the consumption of molecular oxygen in a 600 µl reaction vessel (Model 203B; Instech Labs, Plymouth Meeting, PA). Data are collected using a PC equipped with a Keithley DAS-8 data acquisition board (Taunton, MA). The total µmol of O_2 consumed and the rate of change in oxygen concentration, expressed as µmol O_2 consumed per minute, are obtained from the first derivative of $[O_2]$ vs. time using vendor supplied software (INTAKE. EXE, Instech Labs). Typically, the rate of oxygen consumption

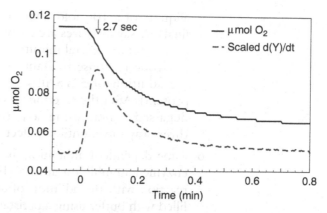

Fig. 3. A representative reaction progress curve for the consumption of dissolved oxygen as a function of time during the cylooxygenase reaction catalyzed by COX, as monitored using the oxygen electrode. The slight lag observed during the approach to maximum rate reflects both activation of the resting enzyme as well as the response time of the electrode. Maximum rate is achieved within approximately 2–3 s following the addition of substrate and is maintained only briefly. The loss of activity observed following the attainment of maximum rate is the result of turnover-dependent enzyme inactivation. The total amount of arachidonic acid provided in the assay is in excess of the concentration of dissolved oxygen and should be sufficient for complete consumption of the O_2 available, in the absence of any enzyme inactivation.

demonstrates a slight lag, and the maximum rate is achieved within 2–4 s after initiating the enzymatic reaction (Fig. 3).

1. Water for rinsing the reaction cell between assays is degassed overnight at room temperature under a mild vacuum to reduce the oxygen concentration and is maintained under a steady flow of nitrogen during the course of the experiment.

2. Oxygraph Assay buffer (item 4 in Subheading 2.1) is prepared and equilibrated at 37°C under air-saturation conditions for at least 1 h prior to use.

3. Stock solutions of purified enzyme (approx. 0.5–5 mg/ml total protein; 7–70 μM COX monomer) are pre-equilibrated with a slight molar excess of hemin, in order to insure reconstitution of the holoenzyme, and are stored on ice.

4. The Clark-style oxygen electrode is prepared daily by electrodeposition of a fresh coating of AgCl and is covered with O_2 probe solution and fitted with a fresh PTFE membrane (Model 5775; YSI, Inc., Yellow Springs, OH), as recommended by the manufacturer. The reaction vessel is assembled and the electrode is allowed to stabilize for 1 h prior to use (see Note 5).

5. Steady state inhibition is examined by initiating the enzymatic reaction with the addition of enzyme stock to a pre-equilibrated reaction mixture containing substrate and test compound. The assay cell is filled with buffer using a peristaltic pump. Arachidonate and test compound are introduced as 6 μl

aliquots of a 100-fold concentrated stock (1% v/v DMSO, final). Assay mixtures are allowed to equilibrate in the reaction vessel for several minutes at 37°C in order to permit the electrode response to stabilize. The reaction is initiated by the addition of 5–15 μl of enzyme stock (6–25 μg total protein). After the reaction is complete, the cell is rinsed with degassed water, using a peristaltic pump attached to a three-way valve, until the electrode response stabilizes.

6. Time-dependent inhibition is examined by pre-incubating enzyme and inhibitor at 37°C before initiating the enzymatic reaction with the addition of arachidonate. The assay cell is filled with buffer using a peristaltic pump, and test compound is introduced as a 6 μl aliquot of a 100-fold concentrated stock (1% v/v DMSO, final). The assay mixture is allowed to equilibrate in the reaction vessel for several minutes at 37°C in order to permit the electrode response to stabilize. An aliquot of enzyme stock, containing 6–25 μg total protein, is added to the reaction mixture containing test compound. After a brief pre-incubation of enzyme with test compound, the reaction is initiated by the addition of 6 μl 10 mM arachidonate (100 μM, final concentration). The pre-incubation period can be varied from several seconds to several minutes in order to monitor for an increase in inhibitory potency as a function of the pre-incubation time. Once the reaction is complete, the cell is rinsed with degassed water, as described above.

3.2. Peroxidase Assay

The peroxidase assay is performed in a 96-well flat bottom microtiter plate in a final volume of 200 μl. The reaction is monitored by following the oxidation of TMPD spectrophotometrically at 610 nm using a microplate reader (SpectraMax Plus, Molecular Devices, Sunnyvale, CA) at room temperature. The peroxidase assay can be utilized as a convenient endpoint assay for the quantitation of enzyme activity from recombinant expression and to monitor activity during protein purification, or as a kinetic assay for the determination of initial rates and the evaluation of time-dependent inhibition by test compounds.

3.2.1. Endpoint Assay

1. Sufficient volumes of Peroxidase Assay buffer (item 5 in Subheading 2.1) and Peroxidase Substrate stock (item 11 in Subheading 2.1) are prepared to support the number of plates to be assayed.

2. Reaction mixtures are constructed by first dispensing 170 μl of assay buffer into wells of a 96-well assay plate.

3. Ten microliters of various dilutions of enzyme stock are added to wells of the assay plate containing buffer. Ten microliters of assay buffer are substituted for enzyme in order to provide a negative control. Assays are typically performed in duplicate or triplicate.

4. The enzymatic reaction is initiated by the addition of 20 µl of Peroxidase Substrate stock containing both arachidonic acid and TMPD (item 11 in Subheading 2.1) to all wells using a multichannel pipettor.

5. The plate is mixed, covered with a plate sealer, and incubated for approximately 1 h, or until the absorbance of the positive controls approaches 1.0 O.D.

3.2.2. Kinetic Assay

In this application, enzymatic reactions are initiated along each vertical column of the 96-well plate separately, rather than simultaneously, in order to vary the pre-incubation time of enzyme with inhibitor. Each column of the assay plate is monitored separately using kinetic mode in order to permit the shortest data collection interval available to the microplate reader (approx. 2 s per data point). A separate assay plate is used for each test compound, constructed as a five-point dose–response curve in each column and replicated across the 96-well plate. Enzyme and inhibitor are pre-incubated for several seconds to minutes and the enzymatic reaction is initiated by the addition of substrate (see Note 6).

1. Sufficient volumes of Peroxidase Assay buffer (item 5 in Subheading 2.1) and Peroxidase Substrate stock (item 11 in Subheading 2.1) are prepared to support the number of plates to be assayed.

2. Compound dilution plates are prepared by dispensing 150 µl of DMSO into wells in rows A through G of a 96-well plate, followed by 225 µl of DMSO into wells in row H. Twenty-five microliters of a 10 mM DMSO stock of each compound to be tested are then added to separate wells in row H and the well contents are mixed. Five 3-fold serial dilutions of test compound are prepared vertically on the plate by successive transfer of 75 µl from wells in row H toward row D, with mixing. Wells in rows A through C are reserved as DMSO blanks to provide a negative and two positive controls. The compound dilution plate is covered with a plate sealer and stored until needed.

3. Separate assay plates are prepared from each compound dilution curve by transferring 10 µl from wells of a single column of the compound dilution plate to corresponding wells in all columns of a clean 96-well flat bottom assay plate.

4. One hundredsixty microliters of assay buffer are added to all wells of each 96-well assay plate.

5. The assay plate is placed in the open plate reader tray and the reader software is programmed to read only one column of the plate at a time in order to permit the shortest data collection interval available to the microplate reader, as recommended by the manufacturer software.

6. Ten microliters of assay buffer are added to the well in row A of the selected column of the assay plate, in order to provide a negative control. Ten microliters of enzyme stock containing approximately 1 μg protein are added to wells in rows B through H using a multichannel pipettor and the well contents are mixed briefly. Enzyme and inhibitor mixtures are allowed to pre-incubate at room temperature for a period of several seconds to several minutes.

7. The enzymatic reactions within the selected column of the plate are initiated by the addition of 20 μl of Peroxidase Substrate stock (item 11 in Subheading 2.1) and the well contents are mixed briefly by repeated withdrawal and dispensing using a multichannel pipettor. Data collection is initiated immediately after mixing.

8. The progress of the enzymatic reaction is monitored continuously in kinetic mode, collecting data points every 2 s for approximately 2 min. Initial rates are determined from the linear portion of the reaction progress curves. Maximum rates of TMPD oxidation are referenced to that of the positive and negative control assays within the respective column on the assay plate, and are expressed as percent activity remaining.

3.3. PGE$_2$ ELISA Assay

The enzymatic assay is performed in a 96-well round bottom microtiter plate in a final volume of 200 μl containing inhibitor and 5% DMSO (v/v final). The enzyme is typically pre-incubated with inhibitor for 10 min at room temperature prior to initiating the enzymatic reaction by the addition of 10 μM arachidonic acid (final concentration). Aliquots of the reaction mixtures are removed after 2 min and 10 min, and are diluted into stop buffer containing 25 μM indomethacin, which serves to quench enzymatic activity. The total amount of PGE$_2$ present in the stopped reaction mixture is quantitated by PGE$_2$ ELISA.

1. Test compounds are initially evaluated using a seven-point dose curve spanning a concentration range from 100 μM to 0.137 μM (final concentration in the assay). Three-fold serial dilutions of test compound are prepared into DMSO along a vertical column of a 96-well round bottom plate. The plate is covered with a sealer until needed.

2. Either purified enzyme or a crude homogenate of insect cells expressing COX-1 or COX-2, containing 1% CHAPS, may be used for analysis of enzyme activity. Purified enzyme stock is diluted to a final conc. 0.5–5 μg/ml protein with assay buffer containing 50 mM potassium phosphate, pH 7.5, 2 μM phenol, and 1 μM hemin immediately prior to use.

3. Enzyme assays are constructed by transferring 10 μl aliquots of test compound from the compound dilution plate into wells of a 96-well round bottom assay plate, in duplicate. One hundred

seventy microliters of the diluted enzyme are added to wells containing inhibitor using a multichannel pipette. Enzyme and inhibitor are pre-incubated at room temperature for 10 min prior to initiating the enzymatic reaction by the addition of 20 µl of 100 µM arachidonic acid (10 µM final concentration), with mixing. The progress of the enzymatic reactions is sampled after 2 and 10 min by transferring 10 or 20 µl aliquots of the reaction mixtures from the assay plate to corresponding wells of a 96-well stopping plate containing 190 or 180 µl, respectively, of 25 µM indomethacin in EIA buffer. Positive and negative controls are provided on each plate by including wells containing DMSO without inhibitor (positive control) and wells containing DMSO but without arachidonate (negative control).

4. After completion of the enzymatic assays, the amount of PGE_2 produced during the reaction is determined by EIA. Blocked EIA plates coated with Goat anti-mouse secondary antibody are washed three times with wash buffer and blotted dry. Twenty microliters of diluted reaction mixture are transferred from wells of the stop plate to corresponding wells of the EIA plate containing 30 µl of EIA buffer (50 µl final).

5. A PGE_2 standard curve is prepared by diluting 40 µl of the 500 ng/ml working stock of PGE_2 in H_2O into 960 µl EIA buffer, for a final concentration of 20 ng/ml PGE_2 in EIA buffer. Ten 2-fold serial dilutions, from 20 ng/ml to 0.039 ng/ml PGE_2, are constructed by successive transfer of 500 µl of each preceding stock into 500 µl EIA buffer in Bio-Rad 960 Titertube Micro Tubes. The serial dilutions can be stored at 4°C for no more than 1 week.

6. The standard curve is constructed on each EIA plate by transferring 50 µl aliquots of the PGE_2 serial dilutions from BioRad Micro Tubes to the top row of the EIA plate using a multichannel pipet. Negative and positive controls are included on each EIA plate by reserving wells A1 and A12 for tracer-only (negative control) and tracer-plus-primary antibody (positive control), respectively.

7. Fifty microliters of PGE_2-AChE tracer are added to all wells of the EIA plate, followed by the addition of 50 µl mouse anti-PGE_2 monoclonal antibody to all wells except well A1. The EIA plate is sealed and incubated in the dark at room temperature overnight.

8. After overnight incubation, the EIA plate is washed three times with EIA buffer and blotted dry. Two hundred microliters of Ellman's Reagent is added to each well and the plate is sealed and incubated at 37°C for several hours to allow color development. Plates are read at 405 nm once the negative controls reach an OD of 0.8–1.0.

3.4. Calculations

For the analysis of steady state inhibition, maximum rates of O_2 consumption obtained from the oxygraph assay are utilized for nonlinear least squares fitting to mathematical models for competitive, noncompetitive, or mixed inhibition using the program GraFit (20). The K_i values obtained represent apparent K_i values determined under conditions of air-saturated aqueous concentrations of oxygen at 37°C.

For the analysis of time-dependent inhibition, maximum rates of O_2 consumption or TMPD oxidation are expressed as percent activity remaining, relative to that of control assays performed in the absence of inhibitor. Irreversible time-dependent inhibition is evaluated based upon a two step model of a reversible second-order reaction followed by an irreversible first-order reaction (18, 21–24). Relative rates of O_2 consumption obtained from the oxygraph assay are utilized for global fitting to the appropriate equation for analysis of irreversible or reversible time-dependent inhibition using the program GraFit. Values for K_m and K_i obtained from steady state kinetic measurements are provided as constants.

Absorbance readings obtained from the PGE_2 ELISA are used to calculate the concentration of PGE_2 generated during the enzymatic reaction by comparison to the PGE_2 standard curve using software provided by the vendor (SoftMax Pro; Molecular Devices, Sunnyvale, CA). The amount of PGE_2 produced is expressed as relative percent with respect to positive and negative controls. Dose–response curves are constructed by plotting relative percent as a function of inhibitor concentration and the data are fit to a four parameter equation for the determination of IC_{50}.

4. Notes

1. Air-saturated water was prepared by routinely storing several liters of purified water on the bench top in an open vessel, loosely capped with a paper wipe or sponge plug, for several days.

2. No precautions were taken to remove or minimize the presence of adventitious peroxides normally present in aqueous stock solutions of arachidonic acid. Trace amounts of lipid peroxides facilitate the activation COX by promoting the formation of the tyrosyl radical required for turnover at the cyclooxygenase active site.

3. TMPD will slowly oxidize in air-saturated aqueous solution. Stocks of TMPD should be kept on ice after thawing and prior to use in order to minimize non-enzymatic oxidation. After use, any remaining volume should be discarded.

4. After several weeks, aqueous stocks of indomethacin demonstrated a loss of potency in the PGE_2 ELISA assay, and the slope

of the resulting IC_{50} curves were observed to decrease. Preparation of a fresh stock of indomethacin generally eliminated this behavior and provided results more consistent with those expected.

5. Overall performance of the oxygen electrode can vary significantly depending upon the source and type of membranes used. The response time and stability of the oxygen electrode is dependent on membrane thickness, and electrode sensitivity will vary slightly each time a new membrane is applied. Of those evaluated, the supplier and model of membranes recommended by the authors provided the best response. Electrode sensitivity can be calibrated using the PES/NADH reaction, as described by Robinson and Cooper (25).

6. The five-point compound dilution curve, with three control wells, is arranged in columns of the assay plate such that all wells in any given row are identical. Enzymatic reactions within each column are initiated separately, by the addition of enzyme to inhibitor, followed by a brief pre-incubation period, and finally by the addition of substrate stock. Therefore, each column of the assay plate represents a separate pre-incubation time point obtained at five concentrations of inhibitor. The plate reader is programmed to read only one column at a time, and a single plate may be utilized to collect up to 12 pre-incubation time points.

References

1. Vane JR (1974) Mode of action of aspirin and similar compounds. In: Robinson JH, Vane JR (eds) Prostaglandin synthase inhibitors. Raven, New York, NY, pp 55–163

2. Smith WL, Marnett LJ (1991) Prostaglandin endoperoxide synthase: structure and catalysis. Biochim Biophys Acta 1083:1–17

3. Smith WL, Garavito RM, DeWitt DL (1996) Prostaglandin endoperoxide H synthases (cyclooxygenases)-1 and -2. J Biol Chem 271:33157–33160

4. DuBois RN, Abramson SB, Crofford L, Gupta RA, Simon LS, van de Putte LBA, Lipsky PE (1998) Cyclooxygenase in biology and disease. FASEB J 12:1063–1073

5. Marnett LJ, Rowlinson SW, Goodwin DC, Kalgutkar AS, Lanzo CA (1999) Arachidonic acid oxygenation by COX-1 and COX-2: mechanisms of catalysis and inhibition. J Biol Chem 274:22903–22906

6. Seibert K, Masferrer JL (1994) Role of inducible cyclooxygenase (COX-2) in inflammation. Receptor 4:17–23

7. Vane JR, Bakhe YS, Botting RM (1998) Cyclooxygenases 1 and 2. Ann Rev Phamacol Toxicol 8:97–120

8. Lecomte M, Laneuville O, Ji C, DeWitt DL, Smith WL (1994) Acetylation of human prostaglandin endoperoxide synthase-2 (cyclooxygenase-2) by aspirin. J Biol Chem 269:13207–13215

9. Gierse JK, Hauser SD, Creely DP, Koboldt C, Rangwala SH, Isakson PC, Seibert K (1995) Expression and selective inhibition of the constitutive and inducible forms of human cyclooxygenase. Biochem J 305:479–484

10. Smith CJ, Zhang Y, Koboldt CM, Muhammad J, Zweifel BS, Shaffer A, Talley JT, Masferrer JL, Seibert K, Isakson PC (1998) Pharmacological analysis of cyclooxygenase-1 in inflammation. Proc Natl Acad Sci USA 95:13313–13318

11. Warner TD, Giuliano F, Vojnovic I, Bukasa A, Mitchell JA, Vane JR (1999) Nonsteroid drug selectivities for cyclo-oxygenase-1 rather than cyclo-oxygenase-2 are associated with human gastrointestinal toxicity: a full in vitro analysis. Proc Natl Acad Sci USA 96:7563–7568

12. Gans KR, Galbraith W, Roman RJ, Haber SB, Kerr JS, Schmidt WK, Smith C, Hewes WE, Ackerman NR (1990) Anti-inflammatory and

safety profile of DuP 697, a novel orally effective prostaglandin synthesis inhibitor. J Pharmacol Exp Ther 254:180–187

13. Futaki N, Takahashi S, Yokoyama M, Aria I, Higuchi S, Otomo S (1994) NS-398, a new anti-inflammatory agent, selectively inhibits prostaglandin G/H synthase/cyclooxygenase (COX-2) activity in vitro. Prostaglandins 47:55–59

14. Penning TD, Talley JJ, Bertenshaw SR, Carter JS, Collins PW, Doctor S, Graneto MJ, Lee LF, Malecha JW, Miyashiro JM, Rogers RS, Rogier DJ, Yu SS, Anderson GD, Burton EG, Cogburn JN, Gregory SA, Koboldt CM, Perkins WE, Seibert K, Veenhuizen AW, Zhang YY, Isakson PC (1997) Synthesis and biological evaluation of the 1, 5-diarylpyrazole class of cyclooxygenase-2 inhibitors: identification of 4-[5-(4-methylphenyl)-3-(trifluoromethyl)-1H-pyrazol-1-yl]benzenesulfonamide (SC-58635, celecoxib). J Med Chem 40:1347–1365

15. Riendeau D, Percival MD, Boyce S, Brideau C, Charleson S, Cromlish W, Ethier D, Evans J, Falgueyret J-P, Ford-Hutchinson AW, Gordon R, Greig G, Gresser M, Guay J, Kargman S, Leger S, Mancini JA, O'Neill G, Oulette M, Rodger IW, Therien M, Wang Z, Webb JK, Wong E, Xu L, Young RN, Zamboni R, Prasit P, Chan C-C (1997) Biochemical and pharmacological profile of a tetrasubstituted furanone as a highly selective COX-2 inhibitor. Br J Pharm 121:105–117

16. Chan C-C, Boyce S, Brideau C, Charleson S, Cromlish W, Ethier D, Evans J, Ford-Hutchinson AW, Forrest MJ, Gauthier JY, Gordon R, Gresser M, Guay J, Kargman S, Kennedy B, Leblanc Y, Leger S, Mancini J, O'Niell GP, Ouellet M, Patrick D, Percival MD, Perrier H, Prasit P, Rodger I, Tagari P, Therien M, Vickers P, Visco

D, Wang Z, Webb J, Wong E, Xu L-J, Young RN, Zamboni R, Riendeau D (1999) Rofecoxib [Vioxx, MK-0966; 4-(4'-methylsulfonylphenyl)-3-phenyl-2-(5H)- furanone]: a potent and orally active cyclooxygenase-2 inhibitor: pharmacological and biochemical profiles. J Pharmacol Exp Ther 290:551–560

17. van der Donk WA, Tsai A-L, Kulmacz RJ (2002) The cyclooxygenase reaction mechanism. Biochemistry 41:15451–15458

18. Gierse JK, Koboldt CM, Walker MC, Seibert K, Isakson PC (1999) Kinetic basis for selective inhibition of cyclo-oxygenases. Biochem J 339:607–614

19. Mnick SJ, Veenhuizen AW, Monahan JB, Sheehan KCF, Lynch KR, Isakson PC, Portanova JP (1995) Characterization of a monoclonal antibody that neutralizes the activity of prostaglandin E2. J Immunol 155:4437–4444

20. Leatherbarrow RJ (1992) GraFit v4.0. Erithacus Software Ltd, Staines, UK

21. Rome LH, Lands WEM (1975) Structural requirements for time-dependent inhibition of prostaglandin biosynthesis by anti-inflammatory drugs. Proc Natl Acad Sci USA 72:4863–4865

22. Kitz R, Wilson IB (1962) Esters of methanesulfonic acid as irreversible inhibitors of acetylcholinesterase. J Biol Chem 237:3245–3249

23. Ouellet M, Percival MD (1995) Effect of inhibitor time-dependency on selectivity towards cyclooxygenase isoforms. Biochem J 306:247–251

24. Huang Z-F, Wun T-C, Broze GJ (1993) Kinetics of factor Xa inhibition by tissue factor pathway inhibitor. J Biol Chem 268:26950–26955

25. Robinson J, Cooper JM (1970) Method of determining oxygen concentrations in biological media, suitable for calibration of the oxygen electrode. Anal Biochem 33:390–399

Part II

Extraction and Measurement of Prostanoids and Isoprostanes

Chapter 12

Extraction and Measurement of Prostanoids and Isoprostanes: Introduction to Part II

Paola Patrignani

Abstract

The assessment of prostanoids (TXB$_2$, 6-keto-PGF$_{1\alpha}$, PGD$_2$ and PGE$_2$) generated in isolated or cultured cells in vitro can be directly assessed in media without extraction mainly using immunoassays, i.e. radio-immunoassay (RIA), enzyme-linked immunosorbent assay (ELISA), or enzyme immunoassay (EIA). However, the immunoassays have to have some fundamental requisites, such as the use of highly specific and sensitive antibodies. Importantly, the results obtained with immunoassays should be validated by comparison with gas chromatography (GC), mass spectrometry (MS), or liquid chromatography (LC)–MS. The measurement of prostanoid generation in vivo can be performed by the evaluation of prostaglandin breakdown products in urine using LC–MS, GC–MS, or validated immunoassays after extraction and rigorous sample purification. The measurement of F$_2$-isoprostanes (F$_2$-iPs), containing F-type ring analogous to PGF$_{2\alpha}$, provides a reliable tool for identifying populations with enhanced rates of lipid peroxidation. Levels of 8- iso-PGF$_{2\alpha}$ (also referred as iPF$_{2\alpha}$-III) in plasma or urine are assessed by GC/MS or validated immunoassays.

Key words: Radioimmunoassays, Immunoassays, Mass spectrometry, Gas chromatography, Liquid chromatography, Isoprostanes, Prostanoids

1. Prostanoids

1.1. Biosynthesis

Prostanoids [i.e. prostaglandin(PG)E$_2$, PGF$_{2\alpha}$, PGD$_2$, prostacyclin(PGI$_2$), and thromboxane(TX)A$_2$] are lipid autacoids which are immediately released outside the cell after intracellular biosynthesis. This occurs probably through the action of MRP4/ABCC4 (multidrug resistance protein 4 adenosine triphosphate-binding cassettes, subfamily C, member 4), an adenosine triphosphate-dependent export pump for organic anions (1).

Prostanoids modulate a wide variety of physiologic and pathologic processes (2). They are generated by three sequential

Samir S. Ayoub et al. (eds.), *Cyclooxygenases: Methods and Protocols*, Methods in Molecular Biology, vol. 644,
DOI 10.1007/978-1-59745-364-6_12, © Springer Science+Business Media, LLC 2010

enzymatic steps involving: (1) phospholipase A_2 enzymes, (2) cyclooxygenase(COX) enzymes(i.e. COX-1 and COX-2) (3), and (3) tissue-specific synthases [3 different enzymes for PGE_2: cytosolic PGE_2 synthase (cPGES), microsomal PGE_2 synthase-1 (mPGES1) and -2 (mPGES2); PGF synthase (PGFS); 2 enzymes for PGD_2:lipocalin-type PGD_2 synthase (L-PGDS) and hematopoietic-type PGD_2 synthase (H-PGDS); PGI_2 synthase (PGIS), thromboxane synthase (TXS)] (4–10). Two isoforms of COX (COX-1 and COX-2) have been cloned and characterized (3). COX-1 is considered a "housekeeping gene" by virtue of constitutive low levels of expression in most cell types. Differently, the gene for COX-2 is a primary response gene with many regulatory sites; thus, COX-2 expression can be rapidly induced by bacterial endotoxin (LPS), cytokines, such as interleukin (IL)-1β and tumor necrosis factor (TNF)-α, growth factors, and tumor promoter phorbol myristate acetate (PMA) (11). However, COX-2 is constitutively expressed in some cells in lung, brain, kidney, pancreatic β-cells, and gastric carcinomas. In endothelial cells, COX-2 is one of the vasoprotective genes upregulated by steady laminar shear stress (LSS) (12) which characterizes "atherosclerotic lesion-protected areas".

1.2. Catabolism

Once generated and released outside the cell, prostanoids are immediately metabolized enzymatically to biologically inactive compounds that are then excreted in urine (13). In addition to the enzymatic metabolism, PGI_2 and TXA_2 are chemically unstable and are rapidly converted to the hydrolysis products, 6-keto-$PGF_{1\alpha}$ and TXB_2, respectively, which are chemically stable but devoid of biological activities. The major catabolic pathway of TXB_2 involves the oxidation of TXB_2 at C-11 catalyzed by NAD^+-dependent 11-hydroxyTXB_2 dehydrogenase (11-TXB_2DH). The product of this reaction, 11-dehydro-TXB_2, has been considered to be a more reliable quantitative index of TX formation in vivo (14). In contrast, the major enzymatic metabolite of prostacyclin and 6-keto-$PGF_{1\alpha}$ in man is 2,3-dinor-6-keto-$PGF_{1\alpha}$ generated through beta-oxidation (15). Moreover, prostaglandins and related eicosanoids are subjected to the oxidation of 15(S)-hydroxyl group catalyzed by NAD^+-dependent 15-hydroxyprostaglandin dehydrogenase (15-PGDH) followed by the reduction of Δ^{13} double bond catalyzed by NADPH/NADH-dependent Δ^{13}-15-ketoprostaglandin reductase (13-PGR). Catabolism of PGE_2 is initiated by 15-PGDH and results in a stable end metabolite, 11-α-hydroxy-9,15-dioxo-2,3,4,5-tetranor-prostane-1,20-dioic acid (PGE-M) that is excreted in the urine and represents an index of systemic PGE_2 production (16). Recently, 11,15-Dioxo-9α-hydroxy-2,3,4,5-tetranorprostan-1,20-dioic acid (tetranor PGDM) has been identified by mass spectrometry as a metabolite of infused PGD_2 that is detectable in mouse and human urine (17). Finally, the main urinary metabolite of $PGF_{2\alpha}$ in humans is 5,7-dihydroxy- 1-ketotetranor-1,16-dioic acid (18).

2. Assessment of Prostanoid Generation

2.1. In Vitro

The assessment of prostanoid generation in isolated or cultured cells is performed by the measurement of the levels of TXB_2, 6-keto-$PGF_{1\alpha}$, PGD_2, and PGE_2 in culture or incubation media. These measurements are performed mainly using immunoassays, i.e. radioimmunoassay (RIA), enzyme-linked immunosorbent assay (ELISA), or enzyme immunoassay (EIA) (19, 20). Prostanoids can be assessed directly in media without extraction. However, the immunoassays have to have some fundamental requisites: (1) the antibodies should not cross-react significantly with other prostanoids that can be present in the medium; (2) the assays should be very sensitive (with detection limits at low picogram range), thus allowing dilution of the sample at 1:10, which avoids the occurrence of possible nonspecific interferences; (3) the results obtained with immunoassays should be validated by comparison with GC–MS. If the levels of prostanoids in media are below the detection limit of the immunoassay or the assessment of prostanoids has to be performed in whole tissues, their extraction from the media is necessary. However, it should be pointed out that all solvents should be accurately eliminated by evaporation because they can interfere with the antibody–antigen reaction of the assays. If the antibodies are not specific for the prostanoid of interest or the prostanoid is a minor product of AA metabolism, it is necessary to separate the prostanoid of interest by HPLC, then collect all fractions corresponding to its retention time, evaporate the solvent, reconstitute with the specific buffer, and assess the concentration by immunoassay (21).

2.2. In Vivo

Arachidonic acid released from biological membranes (and thus the synthesis of prostanoids) is likely to be phasic rather than continuous. This phenomenon also renders the measurement of these products highly liable to artifact when performed in media obtained by invasive sampling techniques, such as plasma. Finally, because of the evanescence of prostanoids in the circulation, there are technical difficulties involved in devising assays of requisite sensitivity and specificity (13).

The measurement of prostaglandin breakdown products in urine can be performed using validated immunoassays or chromatography (GC), mass spectrometry (MS), or liquid chromatography (LC)–MS. GC–MS and LC–MS represents highly specific, noninvasive index of endogenous prostaglandin biosynthesis in vivo. GC–MS is the first physiochemical assay method developed to detect prostanoids in biological fluids. It provides high sensitivity (usually in the low picogram range) and high selectivity based on the mass of analytes or the selected fragments and then GC retention times of analytes. GC allows high accuracy through addition of stable isotope-labeled internal standards to samples (method named isotopic dilution). Limitations of GC–MS are the cost of instrument, the overall complexity of the method, and the requirement for extensive sample purification and derivatization resulting in low throughput. Recently, a new assay that uses the combination of HPLC and MS (LC–MS) which allows profiling of several metabolites of interest present in biological fluids as opposed to the assessment of single components has been developed.

Other advantages of LC–MS versus GC–MS are that the technique is directly applied to substances in solution, it does not require prior derivatization of analytes, it is of moderate complexity, and it provides reproducible MS data. The major limitation of this method is the cost of instrument.

The measurement of urinary metabolites by immunoassays is even more liable to problems of sensitivity and specificity than in plasma. Performance of RIAs in unextracted urine and the use of antibodies with high cross-reactivity to structurally related compounds have been common sources of error. However, the combination of rigorous sample purification with highly specific antibodies has permitted the measurement of urinary prostanoids and their enzymatic metabolites by immunoassays (14, 21).

3. Isoprostanes

Isoprostanes (iPs) are a family of prostaglandin (PG)-like compounds formed nonenzimatically through free radical catalyzed attack on esterified arachidonate followed by enzymatic release from cellular or lipoprotein phospholipids (22). IPs are formed in situ in the phospholipid domain of cell membranes and circulating lipoproteins. Then, they are cleaved by phospholipases, released extracellularly, circulate in plasma, and are excreted in urine. The measurement of F_2-isoprostanes (F_2-iPs), containing F-type ring analogous to $PGF_{2\alpha}$, provides a reliable tool for identifying populations with enhanced rates of lipid peroxidation (23). Levels of 8-iso-$PGF_{2\alpha}$ (also referred as $iPF_{2\alpha}$-III) (24) are most frequently measured in human body fluids, such as plasma or urine. Both stable-isotope dilution assays using gas chromatography/mass spectrometry (GC/MS) and validated radioimmunoassays for 8-iso-$PGF_{2\alpha}$ have been developed (22, 25). The use of immunoassays for the assessment of 8-iso-$PGF_{2\alpha}$ in urine requires rigorous sample purification and the development of highly specific antibodies (25).

References

1. Reid G, Wielinga P, Zelcer N, van der Heijden I, Kuil A, de Haas M, Wijnholds J, Borst P (2003) The human multidrug resistance protein MRP4 functions as a prostaglandin efflux transporter and is inhibited by nonsteroidal antiinflammatory drugs. Proc Natl Acad Sci U S A 100:9108–9110

2. FitzGerald GA (2003) COX-2 and beyond: approaches to prostaglandin inhibition in human disease. Nat Rev Drug Discov 11:879–890

3. Simmons DL, Botting RM, Hla T (2004) Cyclooxygenase isozymes: the biology of prostaglandin synthesis and inhibition. Pharmacol Rev 56:387–437

4. Samuelsson B, Morgenstern R, Jakobsson PJ (2007) Membrane prostaglandin E synthase-1: a novel therapeutic target. Pharmacol Rev 59:207–224

5. Kudo I, Murakami M (2005) Prostaglandin E synthase, a terminal enzyme for prostaglandin E_2 biosynthesis. J Biochem Mol Biol 38:633–638

6. Urade Y, Eguchi N (2002) Lipocalin-type and hematopoietic prostaglandin D synthases as a novel example of functional convergence. Prostaglandins Other Lipid Mediat 68–69:375–382

7. Wu KK, Liou JY (2005) Cellular and molecular biology of prostacyclin synthase. Biochem Biophys Res Commun 338:45–52

8. Ueno N, Takegoshi Y, Kamei D, Kudo I, Murakami M (2005) Coupling between cyclooxygenases and terminal prostanoid synthases. Biochem Biophys Res Commun 338:70–76

9. Komoto J, Yamada T, Watanabe K, Woodward DF, Takusagawa F (2006) Prostaglandin F2alpha formation from prostaglandin H2 by prostaglandin F synthase (PGFS): crystal structure of PGFS containing bimatoprost. Biochemistry 45:1987–1996

10. Haurand M, Ullrich V (1985) Isolation and characterization of thromboxane synthase from human platelets as a cytochrome P-450 enzyme. J Biol Chem 260:15059–15067

11. Kang YJ, Mbonye UR, DeLong CJ, Wada M, Smith WL (2007) Regulation of intracellular cyclooxygenase levels by gene transcription and protein degradation. Prog Lipid Res 46:108–125

12. Topper JN, Cai J, Falb D, Gimbrone MA Jr (1996) Identification of vascular endothelial genes differentially responsive to fluid mechanical stimuli: cyclooxygenase-2, manganese superoxide dismutase, and endothelial cell nitric oxide synthase are selectively up-regulated by steady laminar shear stress. Proc Natl Acad Sci U S A 93:10417–10422

13. FitzGerald GA, Pedersen AK, Patrono C (1983) Analysis of prostacyclin and thromboxane biosynthesis in cardiovascular disease. Circulation 67:1174–1177

14. Ciabattoni G, Maclouf J, Catella F, FitzGerald GA, Patrono C (1987) Radioimmunoassay of 11-dehydrothromboxane B_2 in human plasma and urine. Biochim Biophys Acta 918:293–297

15. FitzGerald GA, Brash AR, Falardeau P, Oates JA (1981) Estimated rate of prostacyclin secretion into the circulation of normal man. J Clin Invest 68:1272–1276

16. Murphey LJ, Williams MK, Sanchez SC, Byrne LM, Csiki I, Oates JA, Johnson DH, Morrow JD (2004) Quantification of the major urinary metabolite of PGE2 by a liquid chromatographic/mass spectrometric assay: determination of cyclooxygenase-specific PGE_2 synthesis in healthy humans and those with lung cancer. Anal Biochem 334:266–275

17. Song WL, Wang M, Ricciotti E, Fries S, Yu Y, Grosser T, Reilly M, Lawson JA, FitzGerald GA (2008) Tetranor PGDM, an abundant urinary metabolite reflects biosynthesis of prostaglandin D2 in mice and humans. J Biol Chem 283:1179–1188

18. Granström E, Samuelsson B (1971) On the metabolism of prostaglandin F2a in female subjects. II. Structures of six metabolites. J Biol Chem 246:7470–7485

19. Patrono C, Ciabattoni G, Pinca E, Pugliese F, Castrucci G, De Salvo A, Satta MA, Peskar BA (1980) Low dose aspirin and inhibition of thromboxane B2 production in healthy subjects. Thromb Res 17:317–327

20. Patrignani P, Panara MR, Greco A, Fusco O, Natoli C, Iacobelli S, Cipollone F, Ganci A, Créminon C, Maclouf J, Patrono C (1994) Biochemical and pharmacological characterization of the cyclooxygenase activity of human blood prostaglandin endoperoxide synthases. J Pharmacol Exp Ther 271:1705–1712

21. Capone ML, Tacconelli S, Sciulli MG, Grana M, Ricciotti E, Minuz P, Di Gregorio P, Merciaro G, Patrono C, Patrignani P (2004) Clinical pharmacology of platelet, monocyte, and vascular cyclooxygenase inhibition by naproxen and low-dose aspirin in healthy subjects. Circulation 109:1468–1471

22. Roberts LJ 2nd, Morrow JD (2002) Products of the isoprostane pathway: unique bioactive

compounds and markers of lipid peroxidation. Cell Mol Life Sci 59:808–820

23. Patrignani P, Panara MR, Tacconelli S, Seta F, Bucciarelli T, Ciabattoni G, Alessandrini P, Mezzetti A, Santini G, Sciulli MG, Cipollone F, Davì G, Gallina P, Bon GB, Patrono C (2000) Effects of vitamin E supplementation on F_2-isoprostane and thromboxane biosynthesis in healthy cigarette smokers. Circulation 102:539–545

24. Rokach J, Khanapure SP, Hwang SW, Adiyaman M, Lawson JA, FitzGerald GA (1997) Nomenclature of isoprostanes: a proposal. Prostaglandins 54:853–873

25. Wang Z, Ciabattoni G, Créminon C, Lawson J, Fitzgerald GA, Patrono C, Maclouf J (1995) Immunological characterization of urinary 8-epi-prostaglandin $F_{2\alpha}$ excretion in man. J Pharmacol Exp Ther 275:94–100

Chapter 13

Prostanoid Extraction and Measurement

Lorenzo Polenzani and Samir S. Ayoub

Abstract

Prostanoids are involved in numerous physiological and pathophysiological processes in all body organs. Therefore, measurement of the concentration of the prostanoids has always been of importance for research and drug development purposes as a measure of cyclooxygenase (COX) activities. Techniques used for the measurement of prostanoids have been described decades ago. These techniques have come a long way and improvements have been reported, especially with the specificity in competition immunoassays that rely on the use of specific antibodies against a given prostanoid. These assays are relatively fast and do not involve the use of radioactive isotopes as radioimmunoassay. However, prior extraction is required in order to concentrate the prostanoids and remove interfering substances such as proteins. In this chapter, we describe two protocols for the extraction and measurement of prostanoids using C18 columns and commercial enzyme immunoassays, which do not require specialized equipments and can be performed in any laboratory with standard equipments.

Key words: C18 columns, Enzyme immunoassay, Extraction, Measurement, Prostaglandin, Prostanoid, Purification

1. Introduction

Prostaglandins (PGs) and thromboxane (TXA_2), which comprise the prostanoids, are arachidonic acid-derived lipid mediators produced from membrane phospholipids, which exert several biological activities as signaling molecules through their cognate G protein-coupled receptors. They are formed by most cells in our body and act as autocrine and paracrine lipid mediators. They are not stored but are synthesized de novo from membrane-released arachidonic acid when cells are activated by physical or chemical stimuli.

The level of each prostanoid as well as their actions are tissue and cell-type specific. There are five major prostanoids that are important in mammalian tissues, prostaglandin E_2 (PGE_2),

Samir S. Ayoub et al. (eds.), *Cyclooxygenases: Methods and Protocols*, Methods in Molecular Biology, vol. 644,
DOI 10.1007/978-1-59745-364-6_13, © Springer Science+Business Media, LLC 2010

prostaglandin $F_{2\alpha}$ ($PGF_{2\alpha}$), prostaglandin D_2 (PGD_2), prostaglandin I_2 (PGI_2), and thromboxane A_2 (TXA_2). They are all synthesized from PGH_2 by nonenzymatic or specific synthases and isomerases. Qualitative and quantitative analysis of prostanoids is a useful index of pharmacological, physiological, and pathological effects (Figs. 1–3) (1–8). Moreover, target opportunity comes from the discovery of a terminal enzyme subset (synthase and isomerases) that address the synthesis to specific final product (9, 10). Thus, the profile changes not only according to the cell type but also depending on the patho-physiological condition and on the drug treatment (11, 12). As a consequence, the method of extraction and quantification of prostanoids must take into account cell and tissue specificity. The extraction conditions are influenced by the tissue used. Relevant variables are for instance the amount of prostanoid in the tissue and the possible presence of contaminants which can interfere with the assay method. Moreover, the low levels

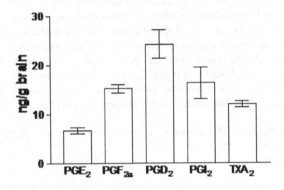

Mean±SE of 6 rat brain

Fig. 1. level of prostanoids obtained from forebrains of rats. The average wet weight of forebrains collected was 1.23 ± 0.071 g ($n = 6$).

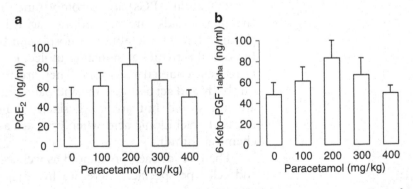

Fig. 2. Prostaglandin E_2 (**a**) and 6-keto-prostaglandin $F_{1\alpha}$ (breakdown product of prostaglandin I_2, **b**) concentrations in peritoneal washouts of mice injected with paracetamol. Peritoneal washouts were collected from mice (Subheading 3.2.1) and prostanoids extracted using C18 columns (Subheading 3.2.2) and prostaglandin E_2 (Subheading 3.3.1) and 6-keto-prostaglandin $F_{1\alpha}$ measured (Subheading 3.3.2).

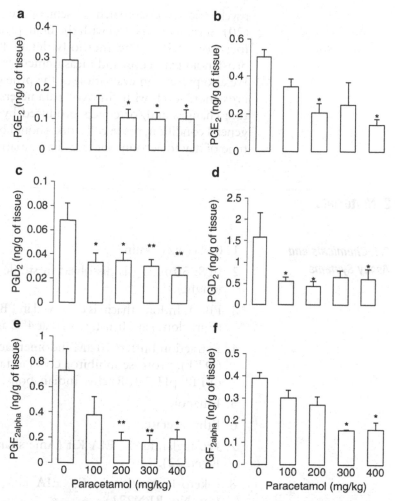

Fig. 3. Prostaglandin E$_2$ (**a** and **b**), prostaglandin D$_2$ (**c** and **d**), and prostaglandin F$_{2\alpha}$ (**e** and **f**) concentrations in spinal cord (**a**, **c**, and **e**) and brain (**b**, **d**, and **f**) tissues of mice injected with paracetamol. Spinal cord and brain tissues were collected from mice (Subheading 3.2.1) and prostanoids extracted using C18 columns (Subheading 3.2.2) and prostaglandin E$_2$ (Subheading 3.3.1), prostaglandin D$_2$ (Subheading 3.3.3) prostaglandin F$_{2\alpha}$ (Subheading 3.3.4) measured.

of prostanoids in the sample may require a procedure suited to concentrate it (13–15). Finally, other fluid or tissue specific factors should also be taken take into account. Of particular relevance is the ex vivo production of prostanoids during procedural steps such as blood drawing. In fact, blood drawing can activate blood cells which are able to synthesize large amounts of prostanoids, quite obviously unrelated to the patho-physiological condition under study.

PGs play important physiological roles in the brain, where they can modulate synaptic transmission, remodeling, neurotransmitter release, hypothalamic–pituitary–adrenal axis, sleep/wake cycle and appetite (16–19). PGs have been linked also to pathological processes in the CNS, including neurodegenerative disorders such as Alzheimer's disease (AD), amyotrophic lateral sclerosis (ALS) and psychiatric disorders such as schizophrenia and mood disorders (20). Several papers deal with the brain prostanoid with a particular focus on PGE_2. The methods detailed in this chapter cover prostanoid extraction and measurement from rodent brain tissues. We also present an evaluation of the protocols used by correlating prostanoid levels with "in vivo" parameters (such as body temperature changes and/or effects produced by drug treatments). As a general condition, tissue collection should be performed at a specific time of day to account for circadian variation.

2. Materials

2.1. Chemicals and Assay Systems

1. 0.9% (w/v) saline.
2. LPS: 50 μg/kg is dissolved in sterile saline (10 ml/kg body weight).
3. PBS-I: Indomethacin is dissolved in PBS without Ca^{2+} and Mg^{2+} (Euroclone) at 10 μM, stored at 4°C and use within 2 days.
4. Extraction buffer: 10 μM indomethacin, 2 mM dithiothreitol (DTT), protease inhibitor cocktail tablets (1 tablet in 25 ml of PBS pH 7.4; Roche, Indianapolis, USA).
5. Ethanol.
6. Ethyl acetate.
7. Prostaglandin E_2 EIA Kit (Amersham Biosciences, Cat. No. RPN222).
8. 6-keto-Prostaglandin $F_{2\alpha}$ EIA kit (Amersham Biosciences, Cat. No, RPN221).
9. Prostaglandin D_2 EIA MOX kit (Cayman Chemicals, Cat No 512011).
10. Prostaglandin $F_{2\alpha}$ EIA kit (Cayman Chemicals, Cat. No. 516011).

2.2. Equipment

1. Nitrogen bomb (Biospec).
2. Centrifuge-Sorvall Legend RT (Jencons).
3. C-18 Sep-Pak columns (Waters).
4. Spin vacuum.
5. Plate reader-GENios Tecan (Jencons).

2.3. Animals

Adult male Wistar rats weighing (200–225 g, Harlan, Italy) and female C57BL/6 mice were maintained under 12 h/12 h light/dark cycle at 22 ± 1°C. Food and water were provided *ad libitum*.

3. Methods

3.1. Prostanoid Extraction Using Extraction Buffer (Fig. 1)

3.1.1. Animals Treatments and Tissue Collection

1. Procedures involving animals and their care were conducted in conformity with the institutional guidelines in compliance with national and international laws and policies.

2. The animals are housed one per polypropylene cage at the time of test. Room temperature and relative humidity are set at $24 \pm 2\,^\circ C$ and $55 \pm 15\%$, respectively, and lighting controlled to give 12 h light and 12 h darkness cycles. Food and water are freely available.

3. Experiments are performed starting between 9:30 and 10:00 a.m. Body temperature is measured by a thermistor rectal probe. Body temperature measured at time (0) is used to assemble homogeneous groups.

4. Animals are sacrificed by exanguination under carbon dioxide unconsciousness.

5. Open the scalp, remove the whole brain and place on ice cooled glass Petri plate covered with bibula paper dipped in PBS-I.

6. Cut the cerebellum off and the remaining forebrain and wash in PBS-I in order to remove remaining blood and inhibit ongoing COX activity.

7. Peel off the choroid plexus and meninges and wash the brain tissues again in PBS-I.

8. Remove excess PBS-I by placing the sample on paper towel.

9. Weigh the tissue, flash-freeze in liquid nitrogen and keep at $-80\,^\circ C$ until use.

3.1.2. Sample Homogenization

1. Pulverize the brain tissue in liquid nitrogen using a mortar and a pestle and suspend in 8 ml of extraction buffer.

2. Keep the samples on ice throughout all the extraction procedure.

3. Homogenize the tissue using a tissue tearer (Ultra-Turrax T25 Janke & Kunkle) 10 s for 3 times in the range of 9,500–13,500/min.

4. Sonicate the homogenate on ice bath for 20 s output 50% (HD 2070 Bandelin Sonoplus) and centrifuge for 30 min at $15,000 \times g$ at $+4\,^\circ C$.

5. Transfer 5 ml of the supernatant to a new tube and split into 1 ml aliquots and store at $-80\,^\circ C$ until analysis.

6. Quantify the remaining supernatant and pellet volume using a graduate plastic pipette.

3.2. Prostanoid Extraction Using C18 Columns (Figs. 2 and 3)

3.2.1. Tissue Collection and Homogenization

1. Kill animals by cervical dislocation.

2. Wash the peritoneal cavity by injecting 1 ml of 0.9% (w/v) saline. Expose the cavity by longitudinal incision of the peritoneal wall and collect the fluid (Fig. 2).

3. Eject the whole spinal cord by initially making incisions at the dorsal and cranial ends of the spinal column and apply pressure using a 23G needle attached to a saline-filled syringe at the dorsal end of the spinal column. Snap-freeze in liquid nitrogen (Fig. 3).

4. Remove the whole brain from the skull and dissect anatomically into the cerebral cortex, mid brain, brain stem, and cerebellum and snap-freeze in liquid nitrogen (Fig. 3).

5. Keep the peritoneal fluid on ice and spin down at 150 g for 10 min at +4°C and store the supernatant at –80°C.

6. Crush spinal cord and brain tissues to powder using a nitrogen bomb previously cooled down in liquid nitrogen. Store powdered tissues at –80°C.

3.2.2. Prostanoid Extraction

1. Suspend powdered spinal cord and brain tissues in 1 ml of 15% (v/v) ethanol (pH 3), leave on ice for 10 min and then spin down at 375 g for 10min at +4°C.

2. Add 500 μl of 15% (v/v) ethanol (pH 3) to 500 μl of peritoneal fluid, leave on ice for 10 min and then spin down at 375 g for 10 min at +4°C.

3. Condition C-18 Sep-Pak columns with 4 ml ethanol followed by 4 ml distilled water at a flow rate of 5–10 ml/min (see Note 1).

4. Apply samples to the columns at a flow rate of 5 ml/min.

5. Wash the columns in 4 ml distilled water followed by 4 ml of 15% (v/v) ethanol (pH 3).

6. Elute the samples with 2 ml of ethyl acetate at a flow rate of 5 ml/min.

7. Evaporate the solvent in a spin vacuum and store at –80°C.

3.3. Prostanoid Measurement Using Enzyme Immunoassay (see Note 2)

3.3.1. Prostaglandin E_2

3.3.1.1. Preparation of Reagents (See Notes 3 and 4)

1. Assay buffer: dilute in 500 ml of distilled water.

2. Wash buffer: dilute in 500 ml of distilled water.

3. Lyophilized anti-PGE_2 antibody: initially dilute in 6 ml assay buffer and then 2 ml of diluted antibody is added to 4ml assay buffer.

4. Lyophilized PGE_2 conjugate: initially dilute in 6 ml assay buffer and then 2 ml of diluted conjugate is added to 4 ml assay buffer.

5. PGE_2 standard: serially dilute as follows: 100 μl of stock standard (256 ng/ml) is diluted in 3.9 ml of assay buffer to give the top standard with concentration of 320 pg/50 μl. Label 7 tubes 2.5, 5, 10, 20, 40, 80, 160 pg/50 μl and add 500 μl

of assay buffer to each. To the 160 pg/50 μl tube add 500 μl of 320 pg/50 μl PGE$_2$ and mix thoroughly. To the 80 pg/50 μl tube, add 500 μl of 160 pg/50 μl and mix thoroughly. Repeat this doubling dilution with the remaining tubes.

3.3.1.2. Assay Procedure

1. Set-up the 96 wells plate coated with goat anti-mouse IgG with sufficient wells to run the standards, controls and samples (see Note 5).

2. Into the nonspecific binding (NSB) wells, add 100 μl of assay buffer.

3. Into the zero standard wells, add 50 μl of assay buffer.

4. Starting with the most dilute add 50 μl of standard PGE$_2$ to the plate in duplicate.

5. Add 50 μl of samples, appropriately diluted, to the plate in duplicate.

6. Add 50 μl of diluted antibody to all the wells except the blank (containing nothing) and NSB.

7. Add 50 μl of diluted conjugate to all the wells except the blank.

8. Cover the plate and incubate at room temperature for 1 h on a microplate shaker.

9. At the end of the 1-h incubation period, wash the plate four times with wash buffer by filling and completely empting the plate between washes.

10. To all the wells, add 150 μl of substrate and incubate the plate at room temperature for 30 min.

11. Read the blue color developed at 630 nm absorbance.

12. Calculate the results by initially calculating the percentage binding (%B/B_0) for standards and samples using the following equation:

$$\%B / B_0 = \frac{(\text{standard} / \text{sample OD} - \text{NSBOD}}{B_0\text{OD} - \text{NSBOD}} \times 100$$

where: B_0 = reading in zero standard wells
OD = Optical density (absorbance values at 630 nm)
NSB = Nonspecific binding wells.

13. Generate a standard curve by plotting the %B/B_0 as a function of the log concentration of PGE$_2$.

14. Read the %B/B_0 for the samples dircetly from the standard curve (see Note 6).

3.3.2. 6-Keto-Prostaglandin $F_{2\alpha}$ (Prostaglandin I_2)

3.3.2.1. Preparation of Reagents (See Notes 3 and 4)

1. Dilute assay buffer in 50 ml of distilled water.

2. Wash buffer: dilute in 500 ml of distilled water.

3. Lyophilized anti-6-Keto-PGF$_{2\alpha}$ antibody: initially diluted in 6 ml assay buffer.

4. Lyophilized 6-Keto-PGF$_{2\alpha}$ conjugate: initially dilute in 6 ml assay buffer.

5. 6-Keto-PGF$_2\alpha$ standard: initially dilute to 1,280 pg/ml with diluted assay buffer. Label seven tubes with 0.5, 1, 2, 4, 8, 16, and 32 and add 500 µl of assay buffer to each. To the 32 pg/50 µl tube, add 500 µl of 1,280 pg/ml 6-Keto-PGF$_{2\alpha}$ and mix thoroughly. To the 16 pg/50 µl tube, add 500 µl of 32 pg/50 µl and mix thoroughly. Repeat this doubling dilution with the remaining tubes.

3.3.2.2. Assay Procedure

1. Set-up the 96 wells plate coated with goat anti-mouse IgG with sufficient wells to run the standards, controls and samples (see Note 5).

2. Into the nonspecific binding (NSB) wells, add 100 µl of assay buffer.

3. Into the zero standard wells, add 50 µl of assay buffer.

4. Starting with the most dilute add 50 µl of standard 6-Keto-PGF$_2\alpha$ to the plate in duplicate.

5. Add 50 µl of samples appropriately diluted to the plate in duplicate.

6. Add 50 µl of diluted antibody to all the wells except the blank (containing nothing) and NSB.

7. Cover the plate and incubate at room temperature for 30 min on a microplate shaker

8. At the end of the incubation, add 500 µl of diluted conjugate to all the wells except the blank.

9. Cover the plate and incubate at room temperature for 1 h on a microplate shaker.

10. At the end of the 1-h incubation period, wash the plate four times with wash buffer by filling and completely empting the plate between washes.

11. To all the wells, add 150 µl of substrate and incubate the plate at room temperature for 30 min.

12. Read the blue color developed at 630 nm absorbance.

13. Calculate the results from this enzyme immunoassay as for PGE$_2$ (Subheading 3.3.1).

3.3.3. Prostaglandin D$_2$

3.3.3.1. Preparation of Reagents (See Notes 3 and 4)

1. Assay buffer: dilute in 90 ml water.

2. Wash buffer: dilute in 2 L of water and add 1 ml of tween-20.

3. PGD$_2$ standard: initially dilute 200 µl of 20 ng/ml Methoximated PGD$_2$ in 600 µl of water to give a concentration of PGD$_2$ of 5 ng/ml. Label eight tubes with 250,125, 62.5, 31.3, 15.6, 7.8, 3.9, 2 pg/ml. To the 250 pg/ml tube, add 950 µl assay buffer and into 125–2 pg/ml tubes add 500 µl of

assay buffer. Transfer fifty microlitres of 5 ng/ml into the 250 pg/ml tube. To the 125 pg/ml tube, add 500 µl of 250 pg/ml PGD_2 mixed thoroughly. To the 62.3 pg/ml tube 500 µl of 125 pg/ml was added and mixed thoroughly. This doubling dilution was repeated with the remaining tubes.

4. PGD_2 antibody: dilute in 6 ml assay buffer.

5. PGD_2 acetylcholinesterase tracer: dilute in 6 ml of assay buffer.

6. PGD_2 methoxime control is used to monitor the degree of the methoximation reaction. Initially dilute 1:10 in water and assay at 1:50 and 1:500.

3.3.3.2. Derivatization of PGD_2 into a Stable Methoxime Form Prior to Measurement

1. Prepare the methyl oximating reagent by initially dissolving methoxylamine HCl in 2–3 ml of a 10:90 ethanol/water solution. Use the methoxylamine HCl solution to dissolve the sodium acetate. Finally, bring the methyoxylamine HCl-sodium acetate solution to a final volume of 10 ml with 10:90 ethanol/water.

2. Derive standard PGD_2 into a stable methoxime form by initially diluting it down 1:10 (100 µl of 400 ng/ml PGD_2 dissolved in 900 µl water to give a concentration of 40 ng/ml) and then heat a 1:1 solution of the standard at 60°C for 30 min with methyl oximating reagent to give a final concentration of standard PGD_2 of 20 ng/ml.

3. Derive the samples into a stable methoxime form by heating 100 µl of sample with 100 µl of methyl oximating reagent at 60°C for 30 min (see Note 7).

3.3.3.3. Enzyme Immunoassay Protocol

1. Set-up the 96 wells plate coated with mouse anti-rabbit IgG with sufficient wells to run the standards, controls and samples (see Note 5).

2. Into the nonspecific binding (NSB) wells, add 100 µl of assay buffer.

3. Into the zero standard wells, add 50 µl of assay buffer.

4. Starting with the most dilute add 50 µl of standard PGD_2 to the plate in duplicate.

5. Add 50 µl of appropriately diluted samples to the plate in duplicate.

6. Add 50 µl of diluted antibody to all the wells except the blank (containing nothing) and NSB.

7. Add 50 µl of diluted acetylcholinesterase tracer to all the wells except the blank.

8. Add 50 µl of PGD_2 methoxime control.

9. Cover the plate and incubate for 18 h at +4°C.

10. Wash the plate five times with wash buffer by filling and completely empting the plate between washes.

11. To all the wells, add $200 \mu l$ of Ellman's reagent (dissolved in 20 ml distilled water). Incubate the plate between 90 and 120 min in the dark with continuous shaking. Read the color developed at 405 nm absorbance.

12. Calculate the results from this enzyme immunoassay as for PGE_2 (Subheading 3.3.1).

3.3.4. Prostaglandin $F_{2\alpha}$

3.3.4.1. Preparation of Reagents (See Notes 3 and 4)

1. Assay buffer: dilute in 90 ml water.

2. Wash buffer: dilute in 2 L water with 1 ml tween-20 added.

3. $PGF_{2\alpha}$ standard: initially dilute $100 \mu l$ of 50 ng/ml $PGF_{2\alpha}$ in $900 \mu l$ of water to give a concentration of $PGF_{2\alpha}$ of 5 ng/ml. Label eight tubes with 250, 125, 62.5, 31.3, 15.6, 7.8, 3.9, 2 pg/ml. To the 250 pg/ml tube add $900 \mu l$ of assay buffer and into 125–2 pg/ml tubes add $500 \mu l$ of assay buffer. Transfer $100 \mu l$ of the 5 ng/ml concentration into the 250 pg/ml tube. To the 125 pg/ml tube add $500 \mu l$ of 250 pg/ml $PGF_{2\alpha}$ and mix thoroughly. To the 62.3 pg/ml tube add $500 \mu l$ of 125 pg/ml and mix thoroughly. Repeat this doubling dilution with the remaining tubes.

4. $PGF_{2\alpha}$ antibody: dilute in 6 ml assay buffer.

5. $PGF_{2\alpha}$ acetylcholinesterase tracer: dilute in 6 ml of assay buffer.

3.3.4.2. Assay Procedure

1. Set-up the 96 wells plate coated with mouse anti-rabbit IgG with sufficient wells to run the standards, controls and samples (see Note 5).

2. Into the nonspecific binding (NSB) wells add $100 \mu l$ of assay buffer.

3. Into the zero standard wells, add $50 \mu l$ of assay buffer.

4. Starting with the most dilute add $50 \mu l$ of standard $PGF_2\alpha$ to the plate in duplicate.

5. Add $50 \mu l$ of samples appropriately diluted to the plate in duplicate.

6. Add $50 \mu l$ of diluted antibody to all the wells except the blank (containing nothing) and NSB.

7. Add $50 \mu l$ of diluted acetylcholinesterase tracer to all the wells except the blank.

8. Cover the plate and incubate for 18 h at +4°C.

9. Wash the plate five times with wash buffer by filling and completely empting the plate between washes.

10. To all the wells add $200 \mu l$ of Ellman's reagent (dissolved in 20 ml distilled water). Incubate the plate between 90 and 120 min

in the dark with continuous shaking. Read the color developed at 405 nm absorbance.

11. Calculate the results from this enzyme immunoassay as for PGE$_2$ (Subheading 3.3.1).

4. Notes

1. For C18, columns without a reservoir solvents were applied by attaching the columns to a syringe filled with the solvents and consistent pressure is applied. It is difficult to be consistent using this technique and thus variation between samples is very likely. In addition, one can only run up to six columns at a time. A much improved technique is to use columns with reservoirs attached to a manifold (such as the one from Varian Inc.) that attach to a vacuum pump.

2. It is advisable to run a trial EIA with one sample per group assayed at a default dilution of for example 1:10. This gives an indication on the dilution to use when performing the assay on all the samples.

3. The reconstituted assay buffer, wash buffer, conjugate and antiserum are stored at +4°C.

4. All the EIA reagents were equilibrated to room temperature prior to use

5. The standards and samples were assayed in duplicate.

6. Only samples that fall within the 20–80% range of the standard curve were analysed further.

7. After methoximation PGD$_2$ is stored at +4°C and is assayed within a few weeks.

References

1. Ayoub SS et al (2004) Acetaminophen-induced hypothermia in mice is mediated by a prostaglandin endoperoxide synthase 1 gene-derived protein. Proc Natl Acad Sci USA 101:11165–11169

2. Trivedi SG, Newson J, Rajakariar R, Jacques TS, Hannon R, Kanaoka Y, Eguchi N, Colville-Nash P, Gilroy DW (2006) Essential role for hematopoietic prostaglandin D2 synthase in the control of delayed type hypersensitivity. Proc Natl Acad Sci USA 103(13):5179–5184

3. Guay J et al (2004) Carrageenan-induced paw edema in rat elicits a predominant prostaglandin E$_2$ (PGE$_2$) response in the central nervous system associated with the induction of microsomial PGE$_2$ Synthase-1. J Biol Chem 279:24866–24872

4. Gühring H et al (2000) Suppresed injury-induced rice in spinal prostaglandin E$_2$ production and reduced early thermal hyperalgesia in iNOS-deficient mice. J Neurosci 20:6714–6720

5. Reinold H et al (2005) Spinal inflammatory hyperalgesia is mediated by prostaglandin E receptors of the EP2 subtype. J Clin Invest 115:673–679

6. Yaksh TL, Kokotos G, Svensson CI, Stephens D, Kokotos CG, Fitzsimmons B, Hadjipavlou-Litina D, Hua XY, Dennis EA (2006) Systemic and intrathecal effects of a novel series of phospholipase A2 inhibitors on hyperalgesia and spinal prostaglandin E2 release. J Pharmacol Exp Ther 316(1):466–475

7. Inoue W, Matsumura K, Yamagata K, Takemiya T, Shiraki T, Kobayashi S (2002) Brain-specific endothelial induction of prostaglandin E(2) synthesis enzymes and its temporal relation to fever. Neurosci Res 44(1):51–61

8. Imai-Matsumura K, Matsumura K, Terao A, Watanabe Y (2002) Attenuated fever in pregnant rats is associated with blunted syntheses of brain cyclooxygenase-2 and PGE$_2$. Am J Physiol Regul Integr Comp Physiol 283:R1346–R1353

9. Engblom D, Saha S, Engstrom L, Westman M, Audoly LP, Jakobsson PJ, Blomqvist A (2003) Microsomal prostaglandin E synthase-1 is the central switch during immune-induced pyresis. Nat Neurosci 6(11):1137–1138

10. Ikeda-Matsuo Y, Ota A, Fukada T, Uematsu S, Akira S, Sasaki Y (2006) Microsomal prostaglandin E synthase-1 is a critical factor of stroke-reperfusion injury. Proc Natl Acad Sci USA 103(31):11790–11795

11. Ciceri P et al (2002) Pharmacology of celecoxib in rat brain after kainate administration. J Pharmacol Exp Ther 302:846–852

12. Pulichino AM, Rowland S, Wu T, Clark P, Xu D, Mathieu MC, Riendeau D, Audoly LP (2006) Prostacyclin antagonism reduces pain and inflammation in rodent models of hyperalgesia and chronic arthritis. J Pharmacol Exp Ther 319(3):1043–1050

13. Araujo P, Froyland L (2006) Optimisation of an extraction method for the determination of prostaglandin E2 in plasma using experimental design and liquid chromatography tandem mass spectrometry. J Chromatogr B Analyt Technol Biomed Life Sci 830(2):212–217

14. Yue H, Jansen SA, Strauss KI, Borenstein MR, Barbe MF, Rossi LJ, Murphy E (2007) A liquid chromatography/mass spectrometric method for simultaneous analysis of arachidonic acid and its endogenous eicosanoid metabolites prostaglandins, dihydroxyeicosatrienoic acids, hydroxyeicosatetraenoic acids, and epoxyeicosatrienoic acids in rat brain tissue. J Pharm Biomed Anal 43(3):1122–1134

15. Yue H, Strauss KI, Borenstein MR, Barbe MF, Rossi LJ, Jansen SA (2004) Determination of bioactive eicosanoids in brain tissue by a sensitive reversed-phase liquid chromatographic method with fluorescence detection. J Chromatogr B Analyt Technol Biomed Life Sci 803(2):267–277

16. Hétu PO, Riendau D (2005) Ciclooxygenase-2 contributes to constitutive prostanoid production in rat kidney and brain. Biochem J 391:561–566

17. Sapirstein A et al (2005) Cytosolic phospholipase A$_2$α regulates induction of brain ciclooxygenase-2 in a mouse model of inflammation. Am J Physiol Regul Integr Comp Physiol 288:R1774–R1782

18. Yamagata K, Matsumura K, Inoue W, Shiraki T, Suzuki K, Yasuda S, Sugiura H, Cao C, Watanabe Y, Kobayashi S (2001) Coexpression of microsomal-type prostaglandin E synthase with cyclooxygenase-2 in brain endothelial cells of rats during endotoxin-induced fever. J Neurosci 21(8):2669–2677

19. Yoshikawa K, Kita Y, Kishimoto K, Shimizu T (2006) Profiling of eicosanoid production in the rat hippocampus during kainic acid-induced seizure: dual phase regulation and differential involvement of COX-1 and COX-2. J Biol Chem 281(21):14663–14669

20. Choi SH, Langenbach R, Bosetti F (2006) Cyclooxygenase-1 and -2 enzymes differentially regulate the brain upstream NF-kappa B pathway and downstream enzymes involved in prostaglandin biosynthesis. J Neurochem 98(3):801–811

Chapter 14

Measurement of 8-Iso-Prostaglandin F$_{2\alpha}$ in Biological Fluids as a Measure of Lipid Peroxidation

Stefania Tacconelli, Marta L. Capone, and Paola Patrignani

Abstract

Several lines of evidence suggest that reactive oxygen species are implicated in human disease, including atherosclerosis, hypertension, and restenosis after angioplasty. The measurement of F$_2$-isoprostanes (F$_2$-iPs), formed nonenzymatically through free radical catalyzed attack on esterified arachidonate, provides a reliable tool for identifying populations with enhanced rates of lipid peroxidation. Among F$_2$-isoPs, 8-iso-PGF$_{2\alpha}$ (also referred to IPF$_{2\alpha}$-III) and IPF$_{2\alpha}$-VI are the most frequently measured in biological fluids. A variety of methods have been proposed to measure F$_2$-isoprostanes in urine and plasma. Mass spectrometry has been developed for the measurement of both F$_2$-isoprostanes but its use is limited as it is time-consuming and highly expensive. We have developed validated enzyme immunoassay (EIA) and radioimmunoassay (RIA) techniques using highly specific antisera for the measurement of 8-iso-PGF$_{2\alpha}$. In contrast, the commercially available immunoassay kits are limited for their poor specificity. The measurement of specific isoprostanes, such as 8-iso-PGF$_{2\alpha}$, in urine is a reliable, noninvasive index of lipid peroxidation that is of valuable help in dose-finding studies of natural and synthetic antioxidant agents.

Key words: Lipid peroxidation, 8-Iso-PGF$_{2\alpha}$, Gas chromatography/mass spectrometry (GC/MS), Radioimmunoassays, Enzymatic immunoassay

1. Introduction

In vitro studies unequivocally demonstrate that all vascular cells produce reactive oxygen species (ROS) and that ROS mediate diverse physiological functions in these cells (1, 2). ROS are constantly generated in vivo and can cause oxidative damage to DNA, lipid, and protein (1, 3–5). The short half-life of these species makes them ideal signalling molecules, but it also confounds their measurement in complex biological systems. In fact, a variety of methods have been proposed and applied for measurement of oxidative DNA, lipid, and protein damage. However, ex vivo indices of oxidant stress could have questionable veracity in assessing

Samir S. Ayoub et al. (eds.), *Cyclooxygenases: Methods and Protocols*, Methods in Molecular Biology, vol. 644,
DOI 10.1007/978-1-59745-364-6_14, © Springer Science+Business Media, LLC 2010

the actual rate of lipid peroxidation in vivo. A valid biomarker should be a major product of oxidative modification, should be a stable product, and should not be susceptible to artifactual induction or loss during storage.

Isoprostanes (isoPs) are a family of prostaglandin (PG)-like compounds formed nonenzymatically through free radical catalyzed attack on esterified arachidonate followed by enzymatic release from cellular or lipoprotein phospholipids (Fig. 1). Bicyclic PGG_2-like endoperoxides exist as four regioisomers, each of which is comprised of eight racemic diastereomers for a total of 64 compounds. The reduction of the bicyclic PGG_2-like endoperoxides undergoes to rearrangement in vivo to form $PGF_{2\alpha}$-like compounds, which are isomeric to $PGF_{2\alpha}$, and thus they have been termed F_2-isoprostanes (F_2-IsoPs) (5–7). IsoPs are formed in situ in the phospholipid domain of cell membranes and circulating lipoproteins. Then, they are cleaved by phospholipases, released extracellularly, circulate in blood, and are excreted in urine. Among F_2-isoPs, 8-iso-$PGF_{2\alpha}$ (also referred as $IPF_{2\alpha}$-III) (6), and $IPF_{2\alpha}$-VI (8) are the most frequently measured in urine samples. In contrast to $IPF_{2\alpha}$-VI, small amounts of 8-iso-$PGF_{2\alpha}$ have been shown to be generated in platelets and monocytes in vitro through cyclooxygenase COX-1- and COX-2-dependent pathways (9, 10); however, most of its generation in vivo occurs independently of COX activity, as demonstrated in experimental models in which the administration of pro-oxidant compounds is

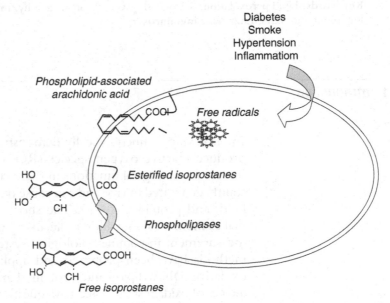

Fig. 1. Isoprostanes are produced in the process of atherothrombosis and vascular inflammation. They are formed nonenzymatically through free radical catalyzed attack on esterified arachidonate followed by enzymatic release from cellular or lipoprotein phospholipids. In the figure, chemical structure of 8-iso-$PGF_{2\alpha}$ is reported.

accompanied by marked increase in 8-iso-$PGF_{2\alpha}$ generation (11). In addition, both urinary F_2-isoPs were unchanged following 2-week dosing with low-dose aspirin (a preferential inhibitor of platelet COX-1) or indomethacin, a nonselective COX inhibitor (8). Interestingly, 8-iso-$PGF_{2\alpha}$, but not $IPF_{2\alpha}$-VI (5, 8), displays biological activity; in fact, it is a vasoconstrictor and modulates platelet activation in response to other agonists. In fact, Praticò et al. (12) found that concentrations of 8-iso-$PGF_{2\alpha}$ ranging from 1 nM to 1 µM induce a dose-dependent increase in platelet morphological changes, increase in calcium release from intracellular store, and in inositol phosphates; it also causes irreversible platelet aggregation, dependent on thromboxane A_2 (TXA_2) generation, when incubated with subthreshold concentrations of ADP, thrombin, collagen, and arachidonic acid. In contrast, higher concentrations of 8-iso-$PGF_{2\alpha}$ (10–20 µM) alone induce weak and reversible platelet aggregation (12). The ability of 8-iso-$PGF_{2\alpha}$ to amplify the aggregation in response to other platelet agonists may be relevant to settings where platelet activation and enhanced free-radical formation coincide (12). Moreover, Minuz et al. (13) found that in the low nM range of concentration, 8-iso-$PGF_{2\alpha}$ and U46619 (a TXA_2 mimetic) triggered shape change and filopodia extension, as well as adhesion to immobilized fibrinogen of washed platelets (13). In addition, using specific inhibitors, it has been found that the signalling pathway triggered by both 8-iso-$PGF_{2\alpha}$ and low concentrations of U46619 to induce platelet adhesion and shape change implicates Syk, the p38 MAP kinase, and actin polymerization (13). These biological effects on platelet function and vascular tone in vivo are mediated via interaction with receptor for TXA_2 (TP) (14). Moreover, proinflammatory effects were identified: 8-iso-$PGF_{2\alpha}$, at 10 µM, induced monocyte adhesion to endothelial cells, in vitro, and increased lipid accumulation in cultured macrophages in the presence of oxidized LDL (15).

Plasma or urine measurement of F_2-isoprostanes appears to provide a sensitive and specific index of nonenzymatic lipid peroxidation and it has been used extensively to assess in vivo lipid peroxidation in several human disease states (5, 7). Interestingly, enhanced generation of 8-iso-$PGF_{2\alpha}$ has been reported in a variety of syndromes associated with oxidant stress: coronary ischemia-reperfusion syndromes, acute coronary syndromes, Alzheimer's disease, chronic obstructive pulmonary disease, and cystic fibrosis; in association with several cardiovascular risk factors, including hypercholesterolemia, hyperhomocysteinemia, diabetes mellitus, renovascular hypertension, and cigarette smoking (16). Thus, measurement of 8-iso-$PGF_{2\alpha}$ provides a reliable tool for identifying populations with enhanced rates of lipid peroxidation (16). A urinary metabolite of 8-iso-$PGF_{2\alpha}$ in humans has been identified by Roberts et al. (17). The major metabolite of 8-iso-PGF$_{2\alpha}$ represents 29% of the total recovered radioactivity infused in humans and it is the product of a single step of β-oxidation and reduction of the Δ^5 double bond, resulting in the formation of 2,3-dinor-5,6-dihydro-8-iso-PGF$_{2\alpha}$. This metabolite exhibits biological properties that are comparable to those of its precursor in retinal and cerebral vasculature as well as on astroglial cells (18). Specifically, 2,3-dinor-5,6-dihydro-8-iso-PGF$_{2\alpha}$ causes constriction of distinct vascular beds with potency similar to that of its precursor, which is largely mediated via TXA$_2$ generation after enhancing calcium entry (18). Thus, 2,3-dinor-5,6-dihydro-8-iso-PGF$_{2\alpha}$ could prolong the biological actions of 8-iso-PGF$_{2\alpha}$ and possibly contribute in sustaining impaired circulation after an oxidant stress (18).

Measurements of the unmetabolized 8-iso-PGF$_{2\alpha}$ can be performed in both plasma and urine; however, most studies have examined its urinary excretion because of the noninvasiveness of the procedure and lack of ex vivo artifactual formation resulting from autooxidation of lipids (see Notes 1–3).

Stable isotope dilution assays using gas chromatography/mass spectrometry (GC/MS) have been developed for 8-iso-PGF$_{2\alpha}$ and IPF$_{2\alpha}$-VI (8, 9). Recently, methods that permit simultaneous analysis of selected members of each of the four classes of F$_2$-isoPs (III–VI) using high performance liquid chromatography/electrospray ionization/tandem mass spectrometry (HPLC/ESI/MS/MS) have been developed (19). Thus, by using homologous internal standards, it is possible to discriminate many individual isomers within a single chromatographic separation, bypassing much of the sample preparation associated with assays based on GC/MS (9). A possible limitation to the use of GC/MS for evaluation of isoprostanes in clinical studies is that GC/MS methods are time-consuming and therefore expensive for large numbers of samples, even if the sensitivity and specificity are excellent (see Notes 4–6). Thus, alternative approaches have been developed and validated to quantify 8-iso-PGF$_{2\alpha}$, i.e., immunoassays (radio- and enzyme-immunoassays, RIA and EIA, respectively) using highly specific antisera for 8-iso-PGF$_{2\alpha}$ (20–22). Moreover, a number of EIA kits have become commercially available, making the measurement of F$_2$-isoprostanes available to many investigators, but with some limitations (see below).

2. Materials

2.1. Solid Phase Extraction and Immunoassays

1. 8-iso-PGF$_2\alpha$ standard (Cayman Chemicals, USA): Stock solutions (1.5 ng/ml) are stored at –80°C.

2. Highly specific antisera are obtained using two different approaches: 8-iso-PGF$_2\alpha$ is conjugated to human serum albumin

using the carbodiimide method for RIA (20) and coupling of 8-iso-PGF$_{2\alpha}$, to keyhole limpet hemocyanin as described earlier (22) for EIA.

3. [^{3}H]-8-iso-PGF$_{2\alpha}$ (15 Ci/mmol) in ethanol solution (Cayman Chemicals USA): store upon arrival at −80°C. 5,6,8,9,11,12,14,15-^{3}H(N)-TXB$_2$ (216 Ci/mmol), (GE Healthcare, UK) stored at −30°C.

4. 5-hydroxy-tempo (stored at 4°C) is light sensitive and is prepared just before using (1 mM final concentration).

5. Bovine serum albumin (Sigma Aldrich, St. Louis, MO). Solution at 5% are stored at −30°C.

6. Solid phase extraction cartridges (C$_{18}$ and silica column) Waters Associates. Vials and scintillation liquid are from Perkin Elmer (USA).

7. Phosphate buffer (0.02 M, pH 7.4): it is prepared by diluting a concentrated solution of 0.25 M.

8. Charcoal (Sigma Aldrich, St. Louis, MO). Suspension (50 mg/ml) is stored at +4°C.

9. Sodium azide (Sigma Aldrich, St. Louis, MO). Solution (0.01%) is stored at +4°C.

2.2. Gas Chromatography (GC) Negative-Ion Chemical Ionization (NICI) Mass Spectrometry (MS)

1. TLC plates: LK6D and 60A Silica Gel Plates (Whatman Inc, NJ).

2. Gas chromatography negative-ion chemical ionization mass spectrometry consists of a Delsi-Nermag Automass 150 (Delsi-Nermag, Argenteuil, France) equipped with a Varian 1077 split/splitless injector operated in the splitless mode, at 260°C. The MS is operated in the negative ion, chemical ionization mode, utilizing methane as the reagent gas. The H$_2$^{18}O (99 atom% ^{18}O) is from Norsk Hydro, Norway.

3. Porcine esterase type (Sigma Aldrich).

4. N-[tert-butyldimethylsilyl]-N-methyltrifluoroacetamide (MTBSTFA; Sigma Aldrich).

2.3. Liquid Chromatography/ Electrospray Ionisation/Tandem Mass Spectrometry (HPLC/ESI/MS/MS)

1. BondElut LMS polymer 100-mg SPE cartridges (Varian).

2. The 0.2-μm Nylon microspin filters (Alltech Associates USA).

3. Sorensen buffer used for solid phase extraction: 1/15 M potassium dihydrogen phosphate and 1/15 M disodium hydrogen phosphate.

4. Methanol-D (99.5 atom% D; Sigma Aldrich).

5. The HPLC system includes an ABI 140B dual syringe pump (Applied Biosystems) and a Hypersil ODS-BD 3 mm, 2.0–3 × 150-mm column (Phenomenex, Torrance, CA).

6. Micromesh Quattro II mass spectrometer (Micromass, USA) equipped with a coaxial electrospray source and triple quadrupole analyzer should be used. The instrument is operated in the negative ion mode with the capillary voltage set to 3 kV, the sampling cone voltage to 40 V, and the source temperature to 70°C. The argon gas pressure in the radiofrequency-only region is set to two microbars. The collision energy is 23 eV. The analyzers are set in the multiple reaction monitoring mode.

3. Methods

3.1. Purification and Extraction for RIA Analysis

1. Overnight urine samples (approximately 8–10 h) are collected in the morning, fractionated in 50 ml aliquots with 5-hydroxy-tempo (5-HT; 1 mM final concentration), a free radical scavenger, and stored at −70°C until analysis. In the case of plasma, 0.005% (v/v) BHT is added prior of lipid extraction (10, 21).

2. After adjusting the pH to 4–4.5 with formic acid, 10 ml urine aliquots are extracted using Sep-Pak C18 cartridges (Waters Associates, USA). Firstly, Sep-Pak C18 cartridges are conditioned with 10 ml of methanol, and then equilibrated with 10 ml of water. Thus, urine sample is loaded onto the cartridge. Then, the cartridge loaded with the urine sample is washed with 10 ml of water and subsequently with 10 ml of hexane to remove interferences, while retaining the analyte. An aliquot of 10 ml of ethyl acetate is used to elute the analyte, retaining more strongly bound interference. The eluate is vacuum-dried in a Speed vac evaporator linked with a Savant-refrigerated condensation trap. After drying, the eluate is reconstituted with 200 μl of benzene/ethyl acetate/methanol (60/40/10, vol/vol) and 800 μl of benzene/ethyl acetate (60/40, vol/vol).

3. The reconstituted eluates are subjected to silica column [conditioned and equilibrated with 5 ml benzene/ethyl acetate (60/40/30, vol/vol) and 5 ml benzene/ethyl acetate/methanol (60/40/10, vol/vol)]. After loading the reconstituted eluates, silica column is washed with benzene/ethyl acetate/methanol (60/40, vol/vol) and eluted with benzene/ethyl acetate/methanol (60/40/30, vol/vol). These eluates are dried with a speed vac system and then reconstituted with 5 ml of phosphate buffer (0.02 M, pH 7.4) and assayed in the RIA system at a final dilution ranging between 1:10 and 1:100.

4. Recovery is evaluated by adding 3,000 cpm of [^{3}H]-TXB$_2$ to each urine sample before extraction and by counting two 1 ml aliquot of the extracted and purified mixture.

5. For RIA measurement, approximately 2,500 cpm of $[^3H]$-8-iso-$PGF_{2\alpha}$ is added to appropriately diluted antiserum in a final volume of 1.5 ml of buffer (phosphate 0.02 M, pH 7.4 and incubated for 72 h at 4°C) to obtain approximately 40–45% binding of the labeled hapten. Separation of antibody bound from free $[^3H]$-8-iso-$PGF_{2\alpha}$, is achieved by rapidly adding 0.1 ml of a 5% bovine serum albumin solution and 0.1 ml of a charcoal suspension (50 mg/ml) and subsequent centrifugation at $3,000 \times g$ for 10 min at 4°C. Supernatant solutions containing antibody-bound PG are decanted directly into 10 ml of scintillation liquid (Optiphase Hisafe II, from Perkin Elmer).

6. Radioactivity is counted in a beta-liquid scintillation counter. Data are processed with the aid of a computer, which is programmed to correct for nonspecific binding.

7. The results are corrected according to creatinine and recovery (calculated from the addition of $[^3H]$-TXB_2 to urinary samples).

3.2. Purification and Extraction for EIA Analysis

1. Purification and extraction of urine for EIA analysis is identical to that performed by Lellouche et al. (23) for prostanoids. Briefly, 8,000 cpm of $[^3H]$-TXB_2 is added to urine for recovery estimation. Ten millilitres of urine aliquots are extracted on Sep-Pak C18 cartridges, previously conditioned with 10 ml of methanol, and then equilibrate with 10 ml of water. Thus, urine sample is loaded onto the cartridge and washed with 10 ml of 2.5% methanol in water. Samples are eluted with 3 ml of methanol/water (80/20, v/v) and evaporated to dryness and then redissolved in 150 μl of methanol/chloroform (1/2, v/v). Then, the extracts are purified on silica gel thin layer chromatography (TLC). Plates are first washed in methanol/chloroform and activated by heating for 1 h at 110°C prior to application of the samples. Samples are developed using chloroform/methanol/acetic acid/water (90:8:1:0.8, v/v) as the solvent. Authentic 8-iso-$PGF_{2\alpha}$ is run on a separate lane and visualized using 3.5% phosphomolybdic acid spray. The zones corresponding to 8-iso-$PGF_{2\alpha}$ (R_f: 0.13) and TXB_2 (R_f: 0.21) are scraped off and eluted with 1 ml EIA buffer [phosphate buffer, 0.1 M, pH 7.4, containing NaCl (0.4 M), EDTA (10^{-3} M), bovine serum albumin (0.1%) and sodium azide (0.01%)].

2. After purification, 8-iso-$PGF_{2\alpha}$ is measured by EIA using a published procedure (21, 22). Tracer (8-iso-$PGF_{2\alpha}$ coupled to acetylcholine esterase from the electric eel) and corresponding antisera are added (50 μl each) at appropriate dilutions on a mouse anti-rabbit IgG coated plate. After incubation for 18 h at room temperature, the plate is developed using

Ellman's reagent. This assay develops in 60–90 min, and then absorbance is read at a wavelength between 405 and 420 nm.

3. Results are expressed in terms of $B/B_0 \times 100$, where B and B_0 are the absorbance of label measured on the bound fraction in the presence or in the absence of 8-iso-PGF$_{2\alpha}$, respectively. Fitting of the standard curves, calculations are done using a linear log-logit transformation.

4. *Preparation of enzymatic tracer*: the carboxylic function of 8-iso-PGF$_{2\alpha}$ is transformed into an activated ester using stoichiometric amounts of N-hydroxysuccinimide (3–6 μmol/ml). Over 90% of the isoprostane can be converted into the corresponding active ester, and the conversion can be verified by TLC analysis using the [^{3}H]-labeled isoprostane as a tracer. When stored anhydrous at –20°C, the active esters are stable for 4–5 months. Then, the ester is subsequently added to the globular G_4 forms of acetylcholine esterase in borate buffer. After 1 h at 22°C, buffer for EIA is added to stop the reaction and to avoid adsorption of the enzyme during subsequent handling. Purification of the conjugates is performed with a BioGel column (90 × 1.5 cm) eluted with Tris buffer, containing sodium azide (21, 22).

3.3. Gas Chromatography/ Negative-Ion Chemical Ionisation/Mass Spectrometry (GC/NICI/MS)

1. 8-iso-PGF$_2\alpha$ is quantified by a stable isotope dilution assay similar to that previously used for TX metabolites (24).

2. The internal standard is [$^{18}O_2$]-8-iso-PGF$_{2\alpha}$ produced from authentic 8-iso-PGF$_{2\alpha}$ using the technique described by Pickett and Murphy (25). Briefly, a phosphate buffer (50 mM, pH 8.0) containing 2 mg of esterase (200 units) is lyophilized, and then reconstituted with 0.25 ml of H$_2$^{18}O. The reaction is initiated by adding the 8-iso-PGF$_2\alpha$ methyl ester dissolved in methanol. The incubation in H$_2$^{18}O is stopped by adding 0.5 ml of methanol, 1.5 ml of ether and 1 ml of H$_2$O. Following a short and vigorous mixing, the ether layer is removed and replaced with an additional 1.5 ml of ether used to further extract the alkaline solution. This extraction quantitatively removes the residual methyl ester of 8-iso-PGF$_{2\alpha}$. After acidification to pH 3.0, the 8-iso-PGF$_{2\alpha}$ free acid is extracted with chloroform. The labeled 8-iso-PGF$_{2\alpha}$ is then purified by silicic acid chromatography. The oxygen-18 content is determined from selected ion monitoring GC–MS of the methyl ester, tris (trimethylsilyl) derivative after methylation of the free acid with diazomethane.

3. Samples are spiked with 1 ng of internal standard and acidified to pH 3–3.5 with formic acid. After extraction on C18 solid phase extraction column, they are purified on TLC plate (LK6D, 60A Silica Gel Plates), using ethyl acetate/methanol/acetic acid (90/10/01%) as mobile phase (9, 21).

4. The samples are derivatized as pentafluorobenzyl (PFB) esters by adding 10 μl of diisopropylethylamine and 20 μl of 10% PFB Br in acetonitrile and allowing the reaction to proceed for at least 10 min at room temperature. The reaction mixture is dried under nitrogen and applied to a further TLC, using ethyl acetate/heptane (80/20%) as mobile phase. The extractions are dried under a stream of nitrogen and the tert-butyldimethylsilyl ether derivative formed by adding 10 μl of N-[tert-butyldimethylsilyl]-N-methyltrifluoroacetamide (MTBSTFA) and 10 μl of pyridine, and allowing the reaction to proceed for 24 h at room temperature. The reaction mixture is then dried under nitrogen, the sample redissolved in 20 μl of dodecane, and analyzed by GC–MS (9, 21).

5. A 30-μm DB-1 capillary column of 0.25 mm inner diameter with 0.25 μm of coating is utilized. The temperature program is from 190 to 320° at 20°C/min. Helium was used as the carrier gas. The interface is maintained at 300°C, the spitless injection temperature is 260°C. The retention time is approximately 14 min. Integration times for selected ion monitoring studies are 500 ms for each of the two ions monitored; m/z 695 for 8-iso-PGF$_{2\alpha}$ and m/z 699 for [^{18}O$_2$]18-iso-PGF$_{2\alpha}$ (9).

3.4. Liquid Chromatography/ Electrospray Ionisation/Tandem Mass Spectrometry (HPLC/ESI/MS/MS)

1. Deuterium exchange of labile hydroxyl and carboxyl protons of the isoP molecules is performed by dissolving 1 μg of isoP in 100 μl of CH$_3$OD, evaporating under a gentle nitrogen stream, and redissolving the sample in 100 μl of CH$_3$OD/D$_2$O (50/50). The deuterium exchanged samples are infused into the electrospray source of the mass spectrometer described below, using CH$_3$OD/D$_2$O (50/50) as the mobile phase (19).

2. Internal standards for F$_2$-isoPs of the four classes are added to 1 ml of urine and allowed to equilibrate for 15 min. Then, 0.5 ml of 15% KOH is added, and the sample is mixed and allowed to stand for 30 min. The pH is then adjusted with 6 M HCl to 2.5. The sample is then subjected to solid phase extraction (SPE). The BondElut LMS polymer 100-mg SPE cartridge is conditioned with 1 ml of ethanol and 1 ml of Sorensen buffer. The sample is applied to the cartridge, which is then sequentially washed with 1 ml each of Sorensen buffer and 5% ethanol in water. The analyte and internal standards are eluted from the cartridge by using 1 ml of ethanol. The eluate is collected and dried under a gentle stream of nitrogen. The resulting residue is then reconstituted with 100 μl of 20%acetonitrile in water and is filtered by centrifugation. The mobile phase in HPLC system consists of water and acetonitrile/methanol (95/5%), both with 0.005% acetic acid adjusted to pH 5.7 with ammonium hydroxide. The flow rate

is controlled at 200 µl/min. The separation is carried out with a linear solvent gradient program starting at 25% B and ramping to 35% B in 20 min. The HPLC eluate is split from 200 µl/min to 50 µl/min before entering the electrospray source.

3. The negatively charged molecular ions of the four classes of iPs undergo extensive collision-induced fragmentation (CID) at a collision energy of 23 eV.

4. The four isoprostanes generate very similar CID fragmentation patterns above m/z 200. However, they showed drastic differences below m/z 200. In fact, an ion at m/z 115 dominates the product ion spectrum of $IPF_{2\alpha}$-VI. Similarly and to a slightly lesser degree, the ion at m/z 193 is the most abundant ion in the 8-iso-$PGF_{2\alpha}$ ($IPF_2\alpha$-III) product ion spectrum. These two ions are also very class specific. The fragment ion at m/z 127 has the highest intensity for $IPF_{2\alpha}$-IV and is also very unique for this class of isomers. The choice for $IPF_{2\alpha}$-V is made based on the specificity of the product ion at m/z 151 (75% of the base peak intensity), even though it is not the base peak in the spectrum (19).

4. Notes

1. For all methods, standardization of urinary isoprostane concentrations by creatinine is required.

2. Isoprostanes have been measured in biological fluids such as urine, plasma, exhaled breath condensate, bronchoalveolar lavage fluid, bile, cerebrospinal, seminal and pericardial fluids. Measurable concentrations of the unmetabolized compound are present in peripheral venous blood, and a relatively reproducible fraction is excreted unchanged in human urine, with no detectable circadian variation (21, 26, 27).

3. It should be emphasized that ex vivo formation of 8-iso-$PGF_{2\alpha}$ can result from autoxidation (25). Immediate processing of the samples and/or adding antioxidant and storage at −70°C appear to reduce this artifactual formation because it did not observe an increase of immunoreactive material up to 6 months under these conditions (26). However, the ex vivo artifactual formation of unmetabolized compounds resulting from autooxidation of lipids is less likely in urine than in plasma.

4. A possible limitation to the use of GC/MS for evaluation of isoprostanes in clinical studies is that GC/MS methods are time-consuming and therefore expensive for large numbers of samples; however, the sensitivity and specificity are excellent.

However, it requires expensive instrumentation, and sample processing is relatively labor intensive.

5. Several alternative mass spectrometric assays have been developed by different investigators, including Morrow et al. (26, 27). Like the assay described (9, 21), these methods require solid phase extraction using a C18 column, TLC purification, and chemical derivatization. Moreover, instead of $[^{18}O_2]$-8-iso-PGF$_2\alpha$ as internal standard (9, 21), $[^2H_4]$-8-iso-PGF$_{2\alpha}$ should be used (26, 27). These assays combine the high resolution of capillary gas chromatography and the selectivity of mass spectrometry, yielding to a picogram sensitivity and a high degree of precision and accuracy (26–28). Many other assays have been developed to improve GC/NICI/MS analysis by using trimethylsilyl ether, instead of the tert-butyldimethylsilyl ether and using methane, isobutane or ammonia as the reactant gases, producing mass spectra poor in mass fragments that are dominated by an intense single ion (29).

6. HPLC/ESI/MS/MS retains the specificity of negative ion chemical ionization GC/MS. Moreover, much of the sample preparation and attendant loss of recovery associated with GC/MS is omitted before HPLC/ESI/MS/MS analysis. The detection limit is 5 pg injected on column requiring a signal-to-noise ratio of ≥ 3. Using IPF$_{2\alpha}$-VI as a model compound, paired analysis of this isoP by HPLC/ESI/MS/MS and GC/NICI/MS gave similar results, which were highly correlated ($r = 0.93$; $P < 0.0001$; $n = 24$) (19).

7. We performed EIA and RIA validation in urine samples, using the following criteria:

   (a) by spiking known amounts of 8-iso-PGF$_{2\alpha}$ (in excess of the concentration found in the biological matrix) to buffer or to two different volumes of urine. There is a good linearity in the quantification of 8-iso-PGF$_{2\alpha}$. The recovery for extraction and further purification of unlabeled 8-iso-PGF$_{2\alpha}$ averaged $73 \pm 9\%$ ($n = 19$) whereas that of $[^3H]$-TXB$_2$ is on average $54 \pm 5\%$ (21);

   (b) by the analysis of increasing volumes of the same pool of urine for which it has been observed a linear relationship between the volume and the assayed material (21);

   (c) by comparison of values obtained by TLC/EIA with an independent analytical approach, GC/NICI/MS. Different samples of urine were analyzed by GC/MS and the same samples were quantified by EIA following purification. An excellent correlation between the two methods was obtained ($r = 0.99$, $n = 9$, $P < 0.001$) (21).

   (d) Additionally, 12 urine samples were extracted and purified for RIA. The urine extracts were quantified by RIA using two antisera with a slight different cross-reactivity. Similar values were obtained using either antiserum ($r=0.99$, $n=12$, $P<0.001$) (21).

8. The lower detection limit for RIA and EIA is 1–2 pg/mL. Moreover, the intra-assay ($n=6$) and inter-assay ($n=8$) coefficients of variation were ±2.0% and ±2.9% at the lowest level of standard (2 pg/ml) and ±3.7% and ±10.8% at the highest level of standard (250 pg/ml), respectively (21). A limitation to the use of RIA is the radioactive consuming and the low specific activity of $[^3H]$-8-iso-PGF$_{2\alpha}$.

9. We have also performed validation of RIA in monocyte culture medium (10): 8-iso-PGF$_{2\alpha}$ levels were measured in monocyte culture medium by direct RIA and by RIA after extraction and HPLC separation, using two antisera with a slight different cross-reactivity (21):

    (a) levels of 8-iso-PGF$_{2\alpha}$ measured by direct RIA were identical to those measured, by both antisera, in HPLC fractions corresponding to the retention time of authentic 8-iso-PGF$_{2\alpha}$ (10);

    (b) spiked known amount of 8-iso-PGF$_{2\alpha}$ in the medium of monocyte was recovered quantitatively when assayed by direct RIA or after extraction and HPLC separation, using both specific antisera (10);

    (c) moreover, correlation between 8-iso-PGF$_{2\alpha}$ levels measured by RIA and EIA in 30 samples of monocyte culture medium was found ($r=0.997$, $P<0.001$) (10).

10. Correlation between 8-iso-PGF$_{2\alpha}$ levels obtained using commercially available EIA kits and GC/MS methods, although significant, was poor ($r=-0.628$, $P<0.02$). Comparison of the methods using the Bland–Altman analysis showed that there were wide limits of agreement between the two methods with only 28% of the values falling within the 95% confidence limits for the difference (30). This poor correlation may be due to the paucity of specificity of the antibody used in the kit, which shows cross-reactivity with 8-iso-PGF$_2\alpha$-like isomers that can be generated from free radical oxidation of arachidonic acid. Thus, reliability of immunoassays is compromised by the possible presence of interfering substances in complex biological fluids.

11. Mass spectrometry has been developed for the measurement of both F$_2$-isoprostanes but its use is limited by time-consuming and high costs. We have developed validated EIA and RIA techniques using highly specific antisera for the measurement of 8-iso-PGF$_{2\alpha}$. In contrast, the commercially available immunoassay kits are limited for their poor specificity. Thus, it is imperative to develop validated methods commercially available to many investigators.

References

1. Griendling KK, FitzGerald GA (2003) Oxidative stress and cardiovascular injury: part I: basic mechanisms and in vivo monitoring of ROS. Circulation 108:1912–1916

2. Griendling KK, FitzGerald GA (2003) Oxidative stress and cardiovascular injury: part II: animal and human studies. Circulation 108:2034–2040

3. Lorch S, Lightfoot R, Ohshima H, Virag L, Chen Q, Hertkorn C, Weiss M, Souza J, Ischiropoulos H, Yermilov V, Pignatelli B, Masuda M, Szabo C (2002) Detection of peroxynitrite-induced protein and DNA modifications. Methods Mol Biol 196:247–275

4. Roberts LJ, Morrow JD (2002) Products of the isoprostane pathway: unique bioactive compounds and markers of lipid peroxidation. Cell Mol Life Sci 59:808–820

5. Patrono C, FitzGerald GA (1997) Isoprostanes: potential markers of oxidant stress in atherothrombotic disease. Arterioscler Thromb Vasc Biol 17:2309–2315

6. Rokach J, Khanapure SP, Hwang SW, Adiyaman M, Lawson JA, FitzGerald GA (1997) Nomenclature of isoprostanes: a proposal. Prostaglandins 54:853–873

7. Patrignani P, Tacconelli S (2005) Isoprostanes and other markers of peroxidation in atherosclerosis. Biomarkers 10:S24–S29

8. Pratico D, Barry OP, Lawson JA, Adiyaman M, Hwang SW, Khanapure SP, Iuliano L, Rokach J, FitzGerald GA (1998) IPF2alpha-I: an index of lipid peroxidation in humans. Proc Natl Acad Sci U S A 95(7):3449–3454

9. Pratico D, Lawson JA, FitzGerald GA (1995) Cyclooxygenase-dependent formation of the isoprostane, 8-epi prostaglandin F2 alpha. J Biol Chem 270:9800–9808

10. Patrignani P, Santini G, Panara MR, Sciulli MG, Greco A, Rotondo MT, di Giamberardino M, Maclouf J, Ciabattoni G, Patrono C (1996) Induction of prostaglandin endoperoxide synthase-2 in human monocytes associated with cyclo-oxygenase-dependent F_2-isoprostane formation. Br J Pharmacol 118:1285–1293

11. Kadiiska MB, Gladen BC, Baird DD, Germolec D, Graham LB, Parker CE, Nyska A et al (2005) Biomarkers of oxidative stress study II: are oxidation products of lipids, proteins, and DNA markers of CCl4 poisoning? Free Radic Biol Med 38:698–710

12. Pratico D, Smyth EM, Violi F, FitzGerald GA (1996) Local amplification of platelet function by 8-Epi prostaglandin F2alpha is not mediated by thromboxane receptor isoforms. J Biol Chem 271(25):14916–14924

13. Minuz P, Andrioli G, Degan M, Gaino S, Ortolani R, Tommasoli R, Zuliani V, Lechi A, Lechi C (1998) The F2-isoprostane 8-epiprostaglandin F2alpha increases platelet adhesion and reduces the antiadhesive and antiaggregatory effects of NO. Arterioscler Thromb Vasc Biol 18(8):1248–1256

14. Audoly LP, Rocca B, Fabre JE, Koller BH, Thomas D, Loeb AL, Coffman TM, FitzGerald GA (2000) Cardiovascular responses to the isoprostanes iPF(2alpha)-III and iPE(2)-III are mediated via the thromboxane A(2) receptor in vivo. Circulation 101:2833–2840

15. Leitinger N, Huber J, Rizza C, Mechtcheriakova D, Bochkov V, Koshelnick Y, Berliner JA et al (2001) The isoprostane 8-iso-$PGF_{2\alpha}$ stimulates endothelial cells to bind monocytes: differences from thromboxane-mediated endothelial activation. FASEB J 15:1254–1256

16. Davi G, Falco A, Patrono C (2004) Determinants of F_2-isoprostane biosynthesis and inhibition in man. Chem Phys Lipids 128:149–163

17. Roberts LJ II, Moore KP, Zackert WE, Oates JA, Morrow JD (1996) Identification of the major urinary metabolite of the F_2-isoprostane 8-iso-prostaglandin F2alpha in humans. J Biol Chem 271(34):20617–20620

18. Hou X, Roberts LJ II, Taber DF, Morrow JD, Kanai K, Gobeil F Jr, Beauchamp MH, Bernier SG, Lepage G, Varma DR, Chemtob S (2001) 2, 3-Dinor-5, 6-dihydro-15-F(2t)-isoprostane: a bioactive prostanoid metabolite. Am J Physiol Regul Integr Comp Physiol 281:391–400

19. Li H, Lawson JA, Reilly M, Adiyaman M, Hwang SW, Rokach J, FitzGerald GA (1999) Quantitative high performance liquid chromatography/tandem mass spectrometric analysis of the four classes of F(2)-isoprostanes in human urine. Proc Natl Acad Sci U S A 96:13381–13386

20. Ciabattoni G (1987) Production of antisera by conventional techniques. In: Patrono C, Peskar BA (eds) Radioimmunoassay in basic and clinical pharmacology, handbook of experimentel pharmacology, vol 82. Springer-Verlag, Berlin, pp 23–68

21. Wang Z, Ciabattoni G, Creminon C, Lawson J, Fitzgerald GA, Patrono C, Maclouf J (1995) Immunological characterization of urinary 8-epi-prostaglandin F2 alpha excretion in man. JPET 275:94–100

22. Pradelles P, Gıssı J, Mıcwur J (1985) Enzyme immunoassay of eicosanoids using acetylcholine esterase as label: an alternative to radioimmunoassay. Anal Chem 57:1170–1173

23. Lellouche F, Fradin A, Fitzgerald G, Maclouf J (1990) Enzyme immunoassay measurement of the urinary metabolites of thromboxane A2 and prostacyclin. Prostaglandins 40:297–310

24. Catella F, FitzGerald GA (1987) Paired analysis of urinary thromboxane B2 metabolites in humans. Thromb Res 47:647–656

25. Pickett WC, Murphy RC (1981) Enzymatic preparation of carboxyl oxygen-18 labeled prostaglandin F2 alpha and utility for quantitative mass spectrometry. Anal Biochem 111:115–121

26. Morrow JD, Hill KE, Burk RF, Nammour TM, Badr KF, Roberts LJ II (1990) A series of prostaglandin F2-like compounds are produced in vivo in humans by a non-cyclooxygenase, free radical-catalyzed mechanism. Proc Natl Acad Sci U S A 87:9383–9387

27. Milne GL, Sanchez SC, Musiek ES, Morrow JD (2007) Quantification of F2-isoprostanes as a biomarker of oxidative stress. Nat Protoc 2(1):221–226

28. Morrow JD (2005) Quantification of isoprostanes as indices of oxidant stress and the risk of atherosclerosis in humans. Arterioscler Thromb Vasc Biol 25:279–286

29. Tsikas DJ (1998) Application of gas chromatography–mass spectrometry and gas chromatography–tandem mass spectrometry to assess in vivo synthesis of prostaglandins, thromboxane, leukotrienes, isoprostanes and related compounds in humans. J Chromatogr B Biomed Sci Appl 717:201–245

30. Proudfoot J, Barden A, Mori TA, Burke V, Croft KD, Beilin LJ, Puddey IB (1999) Measurement of urinary F(2)-isoprostanes as markers of in vivo lipid peroxidation – a comparison of enzyme immunoassay with gas chromatography/mass spectrometry. Anal Biochem 272:209–215

Part III

In Vivo Models to Study Involvement of Cyclooxygenase Products in Health and Disease

Chapter 15

In Vivo Models to Study Cyclooxygenase Products in Health and Disease: Introduction to Part III

Derek W. Gilroy, Melanie Stables, and Justine Newson

Abstract

Inflammation is a primordial response that protects against injury and infection with the ultimate aim of restoring damaged tissue to its normal physiological functioning state. In fact, our well-being and survival depends upon its efficiency and carefully balanced control and to which we are alerted in the form of pain, swelling, and redness. Prostaglandins (PG), lipids derived from arachidonic acid metabolism by the enzyme cyclooxygenase, are historically one of the most well-studied mediators of the acute inflammatory response; so much so that their inhibition by so-called non-steroidal anti-inflammatory drugs (NSAIDs) has been the mainstay for the treatment of diseases where inflammation becomes a pathological driving force. However, while NSAIDs relieve the symptoms of dyregulated inflammatory responses, they do not cure the underlying disease and have associated gastrointestinal and renal toxicity. These side effects arose from inhibiting constitutively expressed, protective cyclooxygenase (COX-1). Finding another inducible COX (COX-2) expressed only at sites of injury provided a new era in inflammation research and a new era in treating inflammation-driven diseases. The hope was that high levels of COX-2 expression at sites of pain and tissue injury drove that disease process and that its inhibition would possess all the benefits of traditional NSAIDS without the side effects that arise from the inhibition of protective COX-1. However, by discussing data derived from experimental disease models of acute inflammation, in this chapter we suggest that delineating between the roles of COX-1 and COX-2 might not be as simple as once thought. We provide examples of data where a pathological role for COX-1 is evident and where COX-2 is clearly protective.

Key words: Inflammatory resolution, Protective lipids, Counter-regulatory pathways

1. Introduction

These are exciting times in inflammation research. Never before has a single, though nonetheless complex, process (inflammation) been implicated in the aetiology of so many apparently diverse diseases ranging from the obvious (arthritis, asthma and psoriasis, for instance) to the no-so clear (cancer, Alzheimer's and heart disease).

Samir S. Ayoub et al. (eds.), *Cyclooxygenases: Methods and Protocols*, Methods in Molecular Biology, vol. 644,
DOI 10.1007/978-1-59745-364-6_15, © Springer Science+Business Media, LLC 2010

And despite our best efforts at dampening inflammation over the years, we have not succeeded at dealing with the problem without paying a hefty price. In the next few chapters, we focus on a pathway arguably investigated, targeted and manipulated more than any other in the history of inflammation research with the ultimate objective of understanding and dampening inflammation-mediated diseases. It is an interesting story that evolved during the early 1990s when a new isoform of an old target, cyclooxygenase, was identified and announced to great fanfare and hope. As a result, inducible cyclooxygenase (COX-2) was billed as the biggest villain in inflammation-driven pathologies in modern times. Sadly, the concept failed through naivety, ignorance and/or a fundamental lack of understanding the aetiology of inflammation and the desire to make money.

2. COX in Health and Disease

The treatment of inflammation-driven diseases began thousands of years ago with the use of salicylate-containing extracts of myrtle, willow bark and poplar trees with salicylate later acetylated in 1897 by Felix Hoffman resulting in the birth of aspirin. Importantly, aspirin and later glucocorticoids formed the mainstay for inflammation therapeutics over the past 100 years or so, their mode of action discerned as being able to inhibit the cardinal signs of inflammation – heat, redness, swelling, and pain. Over the last 40 years, the endogenous signals that cause these symptoms were also elucidated, with aspirin found to inhibit PG (PG) synthesis, factors essential in mediating pain and swelling. Despite these advances, inhibition of COX was not without side effects arising from the protective role arachidonic acid-derived metabolites play in the day-to-day maintenance of tissue homeostasis and health including the GI system. Great hope came when work by Simmons and Herschman identified an inducible isoform of COX (COX-2) delineating to distinct COX isoforms – constitutive COX-1 and inducible COX-2 (1, 2), the hypothesis being that elevated expression of COX-2 at sites of pain and tissue injury drove that disease process and that its inhibition would possess all the benefits of traditional NSAIDS without the side effects that arise from the inhibition of protective COX-1. However, the waters were muddied upon the realization that COX-2 was also protective in the cardiovascular system synthesising vasoactive and anti-platelet PGI_2, whose inhibition of newly marketed COX-2 inhibitors (Vioxx) precipitated heart failure (3). Undoubtedly, the original COX-1 ("good guy") COX-2 ("bad guy") paradigm was a simple and convincing one that captured the imagination of many. However, a number of studies in the years following COX-2's discovery raised questions about the "selective COX-2 inhibitor" theory, particularly with respect to the accuracy of the two principal tenets.

3. COX-2 Inhibition in Inflammation

While COX-2 is highly up-regulated in artificially-stimulated cell culture as well as in intact tissues during infection and injury, COX-1 is also expressed in this setting. The prevailing question at the time asked what the relative expression and temporal profile of COX-1 versus COX-2 through the time course of an experimental inflammation. Moreover, what was the respective contribution of COX-1 versus COX-2 to pathogenic eicosanoid generation responsible for disease symptoms? Remember, central to the "selective COX-2 inhibitor" idea is the assumption that COX-2 is solely responsible for the production of PGs at sites of inflammation. Thus, if COX-1 contributes significantly to the production of PGs, selective blockade of COX-2 will not produce anti-inflammatory effects to the same extent as drugs that inhibit both isoforms. Arising from a flurry of investigation firstly in vitro, then in rodents and finally in man, COX-2 was indeed expressed at sites of inflammation making abundant PG and therefore implicated in driving the symptoms of pain, swelling etc. The latter was deduced using selective COX-2 inhibitors in animal models. For example, the selective COX-2 inhibitor SC58125 reduced pain and inflammation in a rat paw edema model when administered at a dose 100-fold greater than that required to inhibit COX-2-derived PG (4). Given that SC58125 was 100-fold more potent at inhibiting COX-2 over COX-1 *in vitro*, it is likely that in the rat paw oedema model, the effective dose of SC58125 would have also inhibited COX-1. Nonetheless, effects of SC58125 on COX-1 were not measured in that study. In other studies using the rat paw edema model several compounds with a range of *in vitro* selectivity for COX-2 versus COX-1 (NS-398, nimesulide, DuP697, etodolac) were found to significantly reduce inflammation only when given at doses that inhibited COX-1 (5). Suppression of PG synthesis at the site of inflammation correlated with the suppression of COX-1, but not with the suppression of COX-2. At minimally effective anti-inflammatory doses, these compounds significantly inhibited gastric PG synthesis and caused haemorrhagic damage in the stomach that was similar in severity to that seen with an equi-effective anti-inflammatory dose of the dual COX inhibitor, indomethacin (5). Therefore, these results suggest that COX-1 accounts for a significant portion of the PGs produced at a site of inflammation, at least in this model. These findings were subsequently corroborated by Morhan and Langenbach who developed the first COX-1 and COX-2 knockout mice showing that in two models of inflammation, COX-2 deficient mice exhibited inflammatory responses of similar magnitude to those observed in wild-type controls (6), whereas COX-1 knockout mice displayed a dampened inflammatory phenotype compared to wild type controls (7).

4. COX-2 Inhibiton and Gastro-Intestinal Safety

There is abundant evidence that selective inhibitors of COX-2 elicit less gastric damage than conventional nonselective NSAIDs. A number of studies evaluated the effects of drugs with selectivity for COX-2 over COX-1 in animal models of pre-existing gastrointestinal injury and inflammation. The latter is important as most patients taking NSAIDs do not develop clinically significant gastric injury. Instead, it occurs in a subset that is more susceptible to the gastric-damaging actions of these drugs. Therefore, with respect to COX-2 inhibitors, it is more clinically relevant to study these drugs in animal models that more closely resemble the "susceptible" population in humans, i.e., in animals with experimentally induced ongoing GI inflammation. Thus, the majority of PGs produced by the colonic mucosa in a rat model of colitis are derived from COX-2. Daily treatment with a selective inhibitor of COX-2 (L745337) at doses that did not inhibit COX-1 resulted in significant inhibition of mucosal PG synthesis and a marked increase in the severity of colonic damage (8). When treatment was continued for a week, colonic ulceration was exacerbated to the point of perforation of the bowel wall, leading to death in 100% of the rats (8). Therefore, this study directly challenges one of the central tenets of the selective COX-2 hypothesis, namely, that the PGs important for the maintenance of gastrointestinal mucosal integrity are produced solely via COX-1. COX-2 also plays an important role in promoting the healing of ulcers in the stomach. It has been recognized for many years that NSAIDs interfere with the healing of peptic ulcers in humans, whereas administration of PGs accelerates their healing. Mizuno et al. showed that COX-2 was strongly induced in the mouse stomach in which an ulcer had been induced; there was a parallel increase in mucosal PG synthesis (9). Treatment of the mice with the selective COX-2 inhibitor NS-398 resulted in a reduction in mucosal PG synthesis and significant inhibition of ulcer healing (9). These results have recently be extended by Schmassman et al., who showed marked inhibition of gastric ulcer healing in the rat by a COX-2 inhibitor (L745337) (8). Importantly, this inhibition of ulcer healing was observed with doses of L745337 previously shown to produce selective inhibition of COX-2.

Although COX-2 expression in the healthy stomach is low and expression around a site of ulceration is considerably higher, it should be noted that rapid induction of COX-2 can occur, even in response to quite subtle mucosal irritation. Davies et al. demonstrated significant induction of COX-2 in the rat stomach as early as 1 h after administration of aspirin or indomethacin (10). The observation that this rapid up-regulation of COX-2 could be prevented by concomitant administration of a PG suggested that diminished mucosal PG levels were responsible for triggering the compensatory up-regulation of COX-2. In addition, Sawaoka et al. demonstrated marked up-regulation of COX-2 in the rat stomach within 40 min of oral challenge with acid (11). Does this rapid induction of COX-2 have a functional significance? A recent study suggests a crucial role for COX-2 in the so-called "adaptive cytoprotection" response of the stomach, that is, the increase in resistance to injury observed following exposure to a mild irritant. Pretreatment of rats with selective COX-2 inhibitors (NS-398, DFU and L745337) at doses required for anti-inflammatory effects led to inhibition of the defensive response of the stomach to mild irritation. A further functional role for COX-2 in mediating gastric epithelial proliferation has been demonstrated by Sawaoka et al., whereas Nakatsugi et al. have provided evidence that COX-2 contributes to mucosal defense in a rat model of stress ulcer (12).

5. COX-2 and Resolution

Studies on the other end of the inflammatory spectrum, resolution, have revealed a hitherto unappreciated role for COX-2 as well as COX-2/lipoxygenase interaction products (epi-lipoxins, resolvins, etc.) in controlling the severity of inflammation and its longevity. For instance, PGD_2 has emerged recently as an eicosanoid with both pro- as well as anti-inflammatory properties. PGD_2 undergoes dehydration *in vivo* and *in vitro* to yield biologically active PGs of the J_2 series, including PGJ_2, $\Delta^{12,14}$-PGJ_2 and 15-deoxy-$\Delta^{12,14}$-PGJ_2 (15d-PGJ_2). In addition to being a high-affinity natural ligand for anti-inflammatory peroxisome proliferators-activated receptor gamma (PPARγ), 15d-PGJ_2 also exerts its effects through PPARγ-dependent as well as -independent mechanisms to suppress pro-inflammatory signalling pathways and the expression of genes that drive the inflammatory response (13). We have shown that COX-2-derived PGD_2 metabolites contribute to the resolution of acute inflammation (pleuritis) through the preferential synthesis of PGD_2 and 15d-PGJ_2, which, along with the alternative DNA-binding p50–p50 homodimers complexes of NF-κB, bring about resolution by inducing leukocyte apoptosis (14, 15). Indeed, there is an increasing body of evidence detailing the differential effects of PGD_2 metabolites on leukocyte apoptosis as well as the signalling pathways involved. In addition to the well-known eicosanoids of arachidonic acid metabolism, there is a new generation of lipid mediators showing promise as endogenous anti-inflammatories. Resolvins and docosatrienes are fatty acid metabolites of the COX/LOX pathways, where the omega-3 fatty-acid constituents of fish oils docosahexaenoic acid and eicosapentaneoic acid are the substrates and not arachidonic acid. Thus, transcellular metabolism of arachidonic acid by LOX/LOX interaction pathways gives rise to the lipoxin (LX) family of eicosanoid metabolites. LXs display selective actions on leukocytes that include inhibition of PMN chemotaxis, PMN adhesion to and transmigration through endothelial cells, as well as PMN-mediated increases in vascular permeability (16). In contrast to their effects on PMN and eosinophils, LXs are potent stimuli for peripheral blood monocytes, stimulating monocyte chemotaxis and adherence without causing degranulation or release of reactive oxygen species. In fact, LXs and their stable analogues accelerate the resolution of allergic pleural oedema and enhance phagocytosis of apoptotic PMNs by monocyte-derived macrophages in a nonphlogistic fashion, paving the way for a return to tissue normality. LXA_4 and aspirin-triggered 15-epi-LXA_4, as well as their stable analogues, act with high affinity at a G-protein-coupled receptor, LXA_4 receptor (ALXR; also referred to as formyl peptide receptor-like 1 or FPRL1-references). FPRL1 is a member of the family of seven transmembrane G-coupled receptors, which has at least two other members – FPRL2 and the formyl-Met-Leu-Phe receptor (FPR). By contrast, LXB_4 does not bind the ALXR and, although functional studies have indicated the existence of a receptor that is activated by LXB_4, this receptor has not been cloned. As with the cyPGs, the LXs have also been identified as being expressed during and being crucially important for the resolution of acute inflammation. In a model of rat allergic oedema, for instance, LXA_4 was identified along with PGE_2 as being present during the clearance of oedema in this model (17). Inhibition of their synthesis prolonged oedema clearance, which was rescued using stable analogues of these eicosanoids. A recent analysis of eicosanoid synthesis in a murine dorsal air pouch of acute inflammation elicited by TNF-α has revealed a switch in lipid-class metabolism reminiscent of that found in the rat carrageenin-induced pleurisy (18). In response to TNF-α, levels of leukotriene B_4 increased rapidly, followed by PMN infiltration, which coincided with a rise in inflammatory exudate PGE_2. Concomitant with the eventual reduction in PMN numbers and PGE_2 was an increase in LXA_4. It was concluded that PGE_2 induced a switch in lipid mediator synthesis from predominantly 5-LOX-generated leukotriene B_4 to 15-LOX-elicited pro-resolving LXA_4. Along with our findings in the rat carrageenin-induced pleurisy in terms of PG metabolism, this work indicates that, in acute inflammation, lipid-mediator biosynthesis is biphasic, with a role for eicosanoids in the initiation as well as termination of the inflammatory response. Thus, COX-2 seems to play a protective role during inflammation possessing the unique ability to generate counterregulatory eicosanoids. Of course NSAIDS do not distinguish between DDS synthase, and it is likely that they inhibit all PGs. In which case, developing COX downstream synthase inhibitors and knockouts will add refinement to eventually developing drugs that target pathogenic eicosanoids in inflammation and that spare the protective ones.

6. Final Thoughts

COX-2 expression is most evident at sites of inflammation, whereas COX-1 accounts for the majority of PG synthesis in the normal gastrointestinal tract. However, the studies reviewed above suggest that the delineation of the roles of COX-1 and COX-2 might not be quite as clear as has been suggested. Instead, there are distinct examples of circumstances in which COX-2-derived PGs play a crucial role in the maintenance of gastrointestinal mucosal integrity, particularly when the mucosa is ulcerated or inflamed. Alternatively, there are examples in both animal models and clinical studies in which COX-1-derived PGs contribute to the generation of inflammation. Taken together, these findings suggest that selective inhibition of COX-2 might not produce anti-inflammatory effects comparable to those achieved with combined COX-1 and COX-2 blockade. Selective inhibition of COX-2 will result in gastrointestinal toxicity in some circumstances. Moreover, if COX-2 inhibitors have to be given at high doses to achieve desired anti-inflammatory effects, toxicity associated with the inhibition of COX-1 is likely to be encountered. Of course, the ultimate test of the hypothesis that selective COX-2 inhibitors will represent an effective GI-sparing alternative to standard NSAIDs will be the outcome of their use by patients, particularly when used outside the confines of clinical trials. Even if the studies cited above are predictive of significant limitations of COX-2 inhibitors with respect to efficacy and toxicity, it remains possible that these agents will be beneficial for the treatment of some inflammatory conditions.

References

1. Xie WL, Chipman JG, Robertson DL, Erikson RL, Simmons DL (1991) Expression of a mitogen-responsive gene encoding prostaglandin synthase is regulated by mRNA splicing. Proc Natl Acad Sci U S A 88:2692–2696

2. Kujubu DA, Fletcher BS, Varnum BC, Lim RW, Herschman HR (1991) TIS10, a phorbol ester tumor promoter-inducible mRNA from Swiss 3T3 cells, encodes a novel prostaglandin synthase/cyclooxygenase homologue. J Biol Chem 266:12866–12872

3. Grosser T, Fries S, FitzGerald GA (2006) Biological basis for the cardiovascular consequences of COX-2 inhibition: therapeutic challenges and opportunities. J Clin Invest 116:4–15

4. Seibert K, Zhang Y, Leahy K, Hauser S, Masferrer J, Perkins W, Lee L, Isakson P (1994) Pharmacological and biochemical demonstration of the role of cyclooxygenase 2 in inflammation and pain. Proc Natl Acad Sci U S A 91:12013–12017

5. Wallace JL, Bak A, McKnight W, Asfaha S, Sharkey KA, MacNaughton WK (1998) Cyclooxygenase 1 contributes to inflammatory responses in rats and mice: implications for gastrointestinal toxicity. Gastroenterology 115:101–109

6. Morham SG, Langenbach R, Loftin CD, Tiano HF, Vouloumanos N, Jennette JC, Mahler JF, Kluckman KD, Ledford A, Lee CA, Smithies O (1995) Prostaglandin synthase 2 gene disruption causes severe renal pathology in the mouse. Cell 83:473–482

7. Langenbach R, Morham SG, Tiano HF, Loftin CD, Ghanayem BI, Chulada PC, Mahler JF, Lee CA, Goulding EH, Kluckman KD, Kim HS, Smithies O (1995) Prostaglandin synthase 1 gene disruption in mice reduces arachidonic acid-induced inflammation and indomethacin-induced gastric ulceration. Cell 83:483–492

8. Schmassmann A, Peskar BM, Stettler C, Netzer P, Stroff T, Flogerzi B, Halter F (1998) Effects of inhibition of prostaglandin endoperoxide synthase-2 in chronic gastro-intestinal ulcer models in rats. Br J Pharmacol 123:795–804

9. Mizuno H, Sakamoto C, Matsuda K, Wada K, Uchida T, Noguchi H, Akamatsu T, Kasuga M (1997) Induction of cyclooxygenase 2 in gastric mucosal lesions and its inhibition by the specific antagonist delays healing in mice. Gastroenterology 112:387–397

10. Davies NM, Sharkey KA, Asfaha S, Macnaughton WK, Wallace JL (1997) Aspirin causes rapid up-regulation of cyclo-oxygenase-2 expression in the stomach of rats. Aliment Pharmacol Ther 11:1101–1108

11. Sawaoka H, Tsuji S, Tsujii M, Gunawan ES, Nakama A, Takei Y, Nagano K, Matsui H, Kawano S, Hori M (1997) Expression of the cyclooxygenase-2 gene in gastric epithelium. J Clin Gastroenterol 25(Suppl 1):S105–110

12. Nakatsugi S, Terada N, Yoshimura T, Horie Y, Furukawa M (1996) Effects of nimesulide, a preferential cyclooxygenase-2 inhibitor, on carrageenan-induced pleurisy and stress-induced gastric lesions in rats. Prostaglandins Leukot Essent Fatty Acids 55:395–402

13. Straus DS, Glass CK (2001) Cyclopentenone prostaglandins: new insights on biological activities and cellular targets. Med Res Rev 21:185–210

14. Gilroy DW, Colville-Nash PR, Willis D, Chivers J, Paul-Clark MJ, Willoughby DA (1999) Inducible cyclooxygenase may have anti-inflammatory properties. Nat Med 5:698–701

15. Lawrence T, Gilroy DW, Colville-Nash PR, Willoughby DA (2001) Possible new role for NF-kappaB in the resolution of inflammation. Nat Med 7:1291–1297

16. Chiang N, Arita M, Serhan CN (2005) Anti-inflammatory circuitry: lipoxin, aspirin-triggered lipoxins and their receptor ALX. Prostaglandins Leukot Essent Fatty Acids 73:163–177

17. Levy BD, De Sanctis GT, Devchand PR, Kim E, Ackerman K, Schmidt BA, Szczeklik W, Drazen JM, Serhan CN (2002) Multi-pronged inhibition of airway hyper-responsiveness and inflammation by lipoxin A(4). Nat Med 8:1018–1023

18. Levy BD, Clish CB, Schmidt B, Gronert K, Serhan CN (2001) Lipid mediator class switching during acute inflammation: signals in resolution. Nat Immunol 2:612–619

Chapter 16

Protocols to Assess the Gastrointestinal Side Effects Resulting from Inhibition of Cyclo-Oxygenase Isoforms

Brendan J.R. Whittle

Abstract

A prevalent unwanted action of cyclo-oxygenase (COX) inhibitors, as exemplified by the non-steroidal anti-inflammatory drugs (NSAIDs), is their potential to produce gastrointestinal side effects in clinical use. The injury provoked by such agents includes rapid superficial disruption to the surface layer of the gastric mucosa, the production of acute gastric erosions in the corpus region and the formation of ulcers in the antral region of the stomach. The small intestine is also adversely affected, with a developing enteropathy over a more protracted period that causes lesions and inflammation in the gut. From experimental work, the interactive mechanisms of such damage in the stomach differ distinctly from those that underlie the intestinal injury, yet the damage in both regions involves the inhibition of both COX-1 and COX-2 isoforms. This chapter outlines the in vivo methods that can be used to identify the potential for novel NSAIDs and selective COX-inhibitors to produce acute gastric corpus lesions and more-chronic antral ulcers in the rat, as well as causing small intestinal enteropathy. Such methods can also be utilized to evaluate the ability of novel agents to prevent the gastrointestinal injury provoked by NSAIDs or COX-inhibitors.

Key words: Non-steroidal anti-inflammatory drugs, NSAIDs, Gut side effects, Gastric erosions, Antral ulcers, Enteropathy, Small intestinal inflammation, Indomethacin

1. Introduction

It is now well established from both patient trials and retrospective surveys over the past 25 years that chronic use of aspirin and non-steroidal anti-inflammatory drugs (NSAIDs) gives rise to serious iatrogenic effects in the gastrointestinal tract in a significant proportion of the patient population (1). While limited ingestion of such agents can cause widespread but relatively superficial injury to the gastric mucosa, more-chronic use leads to the formation of profound injury, with penetrating ulcers being formed predominantly

Samir S. Ayoub et al. (eds.), *Cyclooxygenases: Methods and Protocols*, Methods in Molecular Biology, vol. 644,
DOI 10.1007/978-1-59745-364-6_16, © Springer Science+Business Media, LLC 2010

in the antral region of humans. Injury to the small intestine can also be provoked by the more-chronic use of NSAIDs in patients (2).

A key mechanism underlying the observed gastric injury involves the inhibition of cyclo-oxygenase (COX) isoforms (3). Experimental studies have indicated that for acute gastric lesions to develop with NSAIDs, inhibition of both COX-1 and COX-2 isoforms is usually necessary (4). Such effects provoke a sequela of events that include the reduction in mucosal blood flow and the endothelial adhesion of pro-inflammatory cells such as neutrophils in the gastric microcirculation (Fig. 1). Other processes, particularly with aspirin and acidic NSAIDs, when administered orally, involve initial topical irritant actions. Such actions disrupt the epithelial continuity, cause mucus and bicarbonate secretory dysfunction, and alter the surface biolayer of zwitterionic phospholipids (5), allowing acid to diffuse back from the lumen into the mucosal tissue. As such topical irritation is greatly accentuated if accompanied by concurrent COX inhibition, this process can assume importance with the oral dosing regimen of NSAIDs in clinical use. NSAIDs also cause injury in the small intestine, with enteropathy being seen with a relatively high prevalence in both patients and experimental models (2, 6) Such injury appears to involve initial inhibition of both COX isoforms, as in the stomach, but has a very different series of events compared with the pathogenesis of gastric mucosal damage (Fig. 2). Experimental studies indicate that subsequent translocation of indigenous bacteria, lipopolysaccharide release, expression of the inducible nitric oxide synthase isoform, iNOS, nitric oxide overproduction and, in combination with the superoxide, the formation of the cytotoxic moiety, peroxynitrite, are key stages of the pathological process (6).

The COX-2 selective agents were originally designed to inhibit prostanoid biosynthesis at inflammatory sites formed by the inducible COX-2 isoform, but not the endogenous protective prostanoids in the gut, thought to be primarily formed by the constitutive

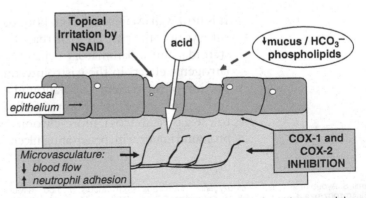

Fig. 1. Interactive mechanisms involved in the pathogenesis of gastric mucosal damage induced by NSAIDs or COX-inhibitors.

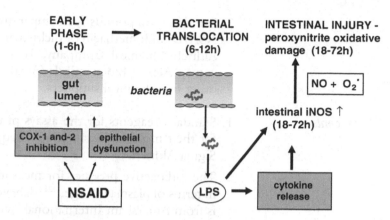

Fig. 2. Interactive mechanisms involved in the pathogenesis of small intestinal enteropathy induced by NSAIDs or COX-inhibitors.

COX-1 (7) and hence avoid gastrointestinal injury, as suggested by earlier pharmacological studies (8). This COX-selectivity concept is now considered somewhat simplistic as it may simply reflect the requirement that inhibition of both isoforms of COX is a prerequisite for the development of overt acute gastric injury (4). Indeed, the lack of gastric injury with the longer-term use of these agents is not fully supported by all clinical data (9–11). It was, however, the increased incidence of adverse cardiovascular events that led to the withdrawal of several of these compounds (12), the first to be removed being rofecoxib (Vioxx).

Thus, the search for novel anti-inflammatory compounds or COX inhibitors, or indeed new pharmaceutical presentation forms of existing compounds that limit gastrointestinal toxicity, still continues. The likelihood is therefore high that animal models of gastrointestinal injury and ulceration such as those developed in the rat which have been used over that past few decades, will continue to be employed. The present chapter describes the key features of rat models that evaluate the ability of NSAIDs and other COX-inhibitors to provoke gastric mucosal damage or small intestinal injury.

2. Materials

2.1. Drugs Under Evaluation

1. Materials for the in vivo study are limited to the NSAIDs, COX-inhibitors or other experimental compounds to be investigated, as well as the vehicle for drug preparation.

2. The vehicle used can be methyl cellulose, carboxymethyl cellulose or more specialised agents such as hydroxypropyl β cyclodextrin (HPCD), which are obtained from the usual general laboratory suppliers such as Sigma Aldrich Chemical Company (www.sigmaaldrich.com).

3. Standard compounds for comparative evaluation such as the NSAIDs indomethacin or diclofenac are obtained from Sigma Aldrich Chemical Company, Tocris Bioscience (www.tocis.com), Alexis Biochemicals/Axxora (www.axxora.com) or other fine chemical suppliers.

2.2. Reagents

1. Standard reagents for the assays of the mediators, or probes for their molecular evaluation if required, are obtained from Sigma Aldrich.

2. The radioactive product for measuring microvascular injury in terms of plasma leakage, ^{125}I-labelled human serum albumin is from Amersham International (www.amersham.com).

3. Methods

3.1. Animal Care and Preparation

1. The animal care and research protocols are in accordance with the local guidelines such as those from the United Kingdom Home Office guidance on the operation of the Animals (Scientific Procedures) Act 1986.

2. Male Wistar rats (200–240 g) are randomised before commencement of the study, housed in grid-bottomed cages in groups of five animals at maximum, at a temperature of 21–25°C with a 12 h light cycle.

3. The animals are inspected and weighed every day. Food and water are provided ad libitum for the period of acclimatisation under the above conditions prior to the study for at least 3 days.

3.2. Acute Gastric Mucosal Lesions

1. For studies on the development of acute gastric mucosal injury, food is withdrawn 18 h prior to the study, but the animals have free access to drinking water

2. The NSAID or selective COX-inhibitor is prepared freshly in the solubilising or suspending vehicle. Some classes of these compounds can be readily prepared in aqueous solution (see Note 1).

3. If not, then the vehicle can be selected from those used routinely such as carboxymethyl cellulose or similar agents (see Subheading 2.1, item 2).

4. The vehicle used should be established not to cause gastric injury itself, or to attenuate the injurious or protective actions of other agents if used in an appropriate concentration (see Note 2).

5. Agents are administered orally via a thin flexible dosing tube inserted down the oesophagus into the stomach, in a standard volume of up to 2 ml/kg or by subcutaneous injection in the same volume.

6. The intravenous route, such as via a tail vein, is rarely used for the administration of such compounds in this model, but may be employed where bioavailability by the other routes is poor or where very rapid COX-inhibition if required for exploratory studies (but see Note 3).

7. The doses of the compounds chosen to be evaluated are those that produce anti-inflammatory effects, or several multiples of such doses, which will allow the estimation of a therapeutic index for the compounds.

8. If appropriate, a dose range of the compound is chosen in order to derive an ED_{50} value (dose causing 50% of the maximum degree of gastric damage).

9. If a standard agent is to be included for comparison, its choice may depend on the structural class of the novel agent being evaluated. However, well-established agents such as the NSAID, indomethacin, are often used (8, 13–18).

10. Using indomethacin, a dose-range of 5–30 mg kg^{-1} by the oral or subcutaneous route is effective in the rat in producing dose-related acute gastric mucosal injury in the acid secreting corpus region, as found with other NSAIDs (8, 13–18).

3.3. Development of Gastric Mucosal Injury

1. Initial lesions in the gastric corpus region are observed within 30 min to 1 h following challenge with the NSAID, particularly if administered orally, but these are more distinct after 3–12 h. Longer periods after challenge, prior to removal of the stomach, may cause increased variability (see Note 4).

2. A development period of 4–6 h after a single dose NSAID administration protocol is used in the routine evaluation of the irritancy of such agents.

3. Multiple dosing of the compound in the study can however be conducted over a 24 h period or longer, should the compound have bio-availability or pharmacokinetic properties that could confound the interpretation after single acute administration. However, issues with using protracted periods of administration need to be considered (see Note 4).

4. After the pre-determined period of challenge, the stomachs are removed for evaluation of the extent of mucosal injury (see Note 5).

3.4. Macroscopic Assessment of Gastric Injury

1. The stomachs are opened along the lesser curvature, from oesophagus to pyloric sphincter, pinned flat and gently washed free of any debris for immediate appraisal of the extent of injury, or photographed for subsequent analysis.

2. Such determinations of mucosal injury are conducted in a randomized and blinded manner.

3. The extent of the mucosal injury can be assessed in quantitative terms by direct measurement of the linear erosions that are macroscopically evident. Using a binocular dissecting microscope equipped with a 0.5–1 mm grid eyepiece, or by computerized planimetry of digital images, the overall length of each erosion is measured and expressed as mm of length. The total length of all the erosions can thus be derived for each stomach.

4. The extent of the injury can also be assessed as a macroscopic score. Each lesion is assigned a 1–3 severity rating, where 0–2 mm length = 1; 2–4 mm length = 2; 4–6 mm length = 3, with the total added score of all the lesions (in some instances, the value is divided by 10) being the erosion score of that stomach.

3.5. Microscopic Assessment of Gastric Injury

1. Samples of gastric mucosal tissue are removed following removal of the stomach for preliminary routine histology.

2. For more detailed histological analysis, rats are terminally anaesthetised with pentobarbitone and perfused via a cannula inserted into the abdominal aorta, with heparinised saline (10 U ml^{-1}) for 5 min, before perfusion fixing with 4% paraformaldehyde for 20 min.

3. Gastric (and intestinal if required) tissues are excised and placed into 4% paraformaldehyde or similar fixative for a further 40 min, before being temporarily stored in phosphate buffered saline.

4. Using standard techniques, tissues are embedded in paraffin wax and frozen. Tissue sections are cut while frozen and dried on a hotplate. Paraffin wax is removed using xylene.

5. The sections are assessed under light microscopy using standard approaches.

3.6. Tissue Preparation for Biochemical or Molecular Measurements

1. For some studies, tissue may be stored for subsequent biochemical assay or evaluation with molecular probes of mediators, including COX products. Whole thickness tissue that includes both muscle and mucosa is removed from the glandular regions of the stomach.

2. It is also possible to separate off the mucosa from the underlying muscle by the use of sharp tweezers, peeling off strips of 1 × 0.5 cm size after lightly cutting through the mucosa with a scalpel.

3. The content of COX products and other mediators are determined in these tissues following appropriate preparation and extraction.

4. To standardise the production of COX products such as prostanoids, PGE_2 or prostacyclin, the tissue strips are chopped with scissors for 15 s in 1 ml volumes of an appropriate buffer in an Eppendorf tube. These are then subjected to vortex-mixing for 1–2 min, followed by rapid centrifugation (15 s at $9,000 \times g$) in a bench centrifuge.

5. This procedure yields a tissue supernatant that can be readily subjected to assay for COX-products and other mediators without extensive extraction.

6. Alternatively, the tissue is homogenised in an appropriate buffer and the supernatant stored after centrifugation for extraction and subsequent assay.

7. The details of the various assays for the prostanoids or the molecular probing of the tissues are beyond the scope of this report and are dealt with elsewhere in this book.

3.7. Assessment of Inhibitors of NSAID-Induced Gastric Injury

1. The procedures described above relate to the induction of acute gastric lesions by standard and experimental NSAIDs and COX-inhibitors. However, the methodology can be readily adapted to evaluate the potential of novel compounds to prevent such gastric damage by known NSAIDs.

2. In such studies, the experimental compound is administered by the chosen route prior to challenge with the standard NSAID, often indomethacin (10–30 mg kg^{-1}p.o. or s.c.). The pre-treatment time will depend of the pharmacokinetic profile of the compound under study, but is usually between 1–3 h, or multiple dosing over 24 h.

3. The degree of gastric injury in the treatment groups are then assessed 3–6 h following NSAID challenge, and compared with that induced in the control challenged group by the NSAID alone (with appropriate controls for the vehicles used).

3.8. Development of Gastric Antral Ulcers by NSAIDs

1. In contrast to the acute, superficial, rapidly healing linear gastric erosions which are provoked by NSAIDs in starved rats, indomethacin and some other NSAIDs produce irregular shaped, penetrating and persistent ulcers in the antral region of the rat gastric mucosa when the animals have been starved and then refed before or soon after challenge with the NSAID (19, 20).

2. This model produces predominantly antral ulcers which develop over 3–12 h, the injury penetrating deep into the mucosa and submucosa to form true ulcers, and which are still evident some 7 days after the single administration of the NSAID.

3. These ulcers thus more closely resemble those observed in the clinic following the more chronic exposure to NSAIDs.

4. The original report (17) specifies that animals should be starved for 24 h, refed for 1 h and then challenged with a single dose of indomethacin (30 mg/kg), subcutaneously (see Note 6).

5. The stomachs are removed at various times after indomethacin challenge, from 5 h to 7 days, opened along the greater curvature and the gastric contents removed for evaluation using a dissecting microscope.

6. The antral lesions are distinct after 5 h, reach their maximal depth though the mucosa and submucosa after 12–24 h (without their surface area necessarily increasing) and are still macroscopically apparent after some 7 days.

7. The lesions and ulcers that develop in the antral region are non-symmetrical and are measured using a calibrated binocular eyepiece or via computerized planimetry of digital images in terms of their area in mm². The total area of ulceration for each stomach is reported in terms of mm².

8. For routine or screening purposes, the model can be used for evaluation of novel compounds, 6 h following challenge, as distinct antral lesions and even penetrating antral ulcers can be observed, depending on the challenging dose of indomethacin (see Note 6).

3.9. Injury to the Small Intestine by NSAIDs

1. NSAIDs will cause injury to the rat small intestine when administered by oral or subcutaneous routes, but the protocol and time-course of development is substantially different to that for provoking gastric injury (21–26).

2. Male Wistar rats (200–250 g) receive both food and water ad libitum during the course of these experiments (see Note 6).

3. As a standard agent for comparison, indomethacin (10 mg kg⁻¹; s.c.) can be selected to produce macroscopic jejunal injury (21–26). Other NSAIDs can likewise be used as standards (see Note 7).

4. These doses of the NSAID can also be used to evaluate the actions of these agents in provoking intestinal injury after their oral, rectal or parenteral administration (24).

5. At various times from 18–72 h, or onwards following drug administration, the small intestine is removed for analysis and measurement of the enteropathy. Standard segments of intestine are taken, opened, pinned flat and gently washed free of debris for quantitative assessment of the area of involvement, as well as an evaluation of the severity of the enteropathy.

3.10. Macroscopic Assessment of Intestinal Damage

1. The extent of intestinal injury is assessed from the area of involvement. This is measured from digital images of the jejunal or ileal mucosa by computerized planimetry and expressed as percentage of the total area of the intestinal segment under study.

2. The severity of the injury and ulceration is also assessed in a randomized blinded fashion on a scale of 1–5, and expressed as a damage index.

3. For the damage index, scoring values are: 0 = no damage to tissue; 1 = the appearance of palpable white nodules and single haemorrhagic segment extending less than 1 cm; 2 = Single zone of mucosal erosion and ulceration extending less than 10 cm with haemorrhage; 3 = Single zone of mucosal erosion and ulceration extending more than 10 cm with haemorrhage; 4 = Multiple zones of mucosal erosion and ulceration with adhesions and luminal bleeding; 5 = Extensive adhesions of bloated abdominal viscera due to perforating lesions; intestinal content black in colour due to blood in lumen.

3.11. Vascular Leakage as an Assessment of Intestinal Damage

1. Albumin leakage using a radioactive marker can also be used as a quantitative marker of vascular injury resulting from the damage to the mucosa in the jejunum and ileum and indeed throughout the gastrointestinal tract and in other organs (26, 27).

2. Under transient halothane anaesthesia, ^{125}I-labelled human serum albumin ($[^{125}I]$-HSA, 2 μCi kg^{-1}, s.c.) is injected via a tail vein, 2 h before autopsy.

3. The leakage of $[^{125}I]$-HSA is subsequently determined in segments of jejunal tissues (3 cm; removed 10–15 cm from the pyloric sphincter) or ileal tissue, both taken from rats terminally anaesthetised with halothane.

4. Blood is collected from the abdominal aorta into syringes, containing trisodium citrate (final concentration 0.318%) and centrifuged (10,000 × g, 10 min, 4°C).

5. The $[^{125}I]$-HSA content of the plasma (100 μl) and in the segments of the tissue is determined in a gamma spectrometer and the albumin content in the tissues is hence calculated.

6. In a separate group of rats, intravascular volume is assessed in terms of the activity of $[^{125}I]$-HSA in tissues after a 2 min circulation.

7. Values from control tissue are subtracted from the values of treated tissue and the data are expressed as plasma leakage, Δ plasma μl g^{-1} tissue, corrected for intravascular volume.

4. Notes

1. Some NSAIDs such as the acidic class exemplified by indomethacin can be readily prepared in weak alkaline solutions or buffers such as 5% sodium bicarbonate solution and once the compound is dissolved, diluted down 1–4 to yield a 1.25%

sodium bicarbonate solution with water-for-injection. This procedure provides a dosage formulation that is close to being an isotonic solution and can thus be readily used for parenteral administration.

2. Certain vehicles or solubilising agents such as ethanol or dimethyl sulphoxide (DMSO) must always be treated with caution, especially for oral administration, as they themselves in high concentrations can exert pharmacological actions on the gut, may interact with some other agents under study and exhibit pro- or anti-ulcer properties. When using such vehicles, a number of control studies are therefore strongly advisable for these agents.

3. While the inhibition of the biosynthesis of COX products in the gastric mucosal tissue by NSAIDs is usually rapid, and can be observed within 10 min following administration by their oral and subcutaneous route, the development of macroscopically apparent gastric injury and erosions is relatively slower.

4. Longer periods than 12 h after challenge with the NSAID can produce variable results as the gastric mucosa will then be undergoing rapid repair of the superficial mucosa by the process of restitution, as also seen following a variety of insults to the gastric mucosa (28).

   Thus, NSAIDs that provoke macroscopically detectible but relatively low degree of damage when assessed at 6 h after a single challenge, may appear to be devoid of such injury if only measured macroscopically after 24 h, for example. Injury at this time may be observed by microscopic and histological evaluation, however. Repeated challenge with the NSAID, for example over 24 h will generally prevent the rapid healing of the mucosa over this period, and will mimic many of the clinical treatment protocols. However, tolerance to the injurious effects can diminish over longer periods, depending on time and frequency of exposure to the NSAID (29).

5. A solution of Evans Blue (1%) can be injected (2 ml kg^{-1} i.v.) 10 min prior to removal of the gastrointestinal organs to help visualise any very minor lesions, if any, although they may not be deemed necessary depending on the degree of tissue injury and the skill of the observer.

6. For the induction of antral ulcers by NSAIDs, variations of this refeeding schedule have been adopted. Thus, administering the indomethacin 1 h prior to refeeding after a 24 h fast gives a higher degree of antral ulceration (19, 20). Other variations to the original method include increasing the dose of indomethacin to 60 mg/kg, which yields deep ulcers at 6 h after challenge (30).

7. In contrast to the development of acute gastric lesions, withdrawal of food is not necessary for the induction of small intestinal lesions and indeed, prevents the development of the enteropathy.

8. As standard agents for use in studies on intestinal lesions, diclofenac (10–40 mg kg^{-1}; s.c.) and flurbiprofen (10–40 mg kg^{-1}; s.c.) provide reproducible levels of macroscopic jejunal injury (24).

Acknowledgments

The author is supported by a grant from the William Harvey Research Foundation, London.

References

1. García Rodríguez LA, Barreales Tolosa L (2007) Risk of upper gastrointestinal complications among users of traditional NSAIDs and COXIBs in the general population. Gastroenterology 132:498–506

2. Laine L, Connors LG, Reicin A, Hawkey CJ, Burgos-Vargas R, Schnitzer TJ, Yu Q, Bombardier C (2003) Serious lower gastrointestinal clinical events with nonselective NSAID or coxib use. Gastroenterology 124:288–292

3. Whittle BJR (2003) Gastrointestinal effects of nonsteroidal anti-inflammatory drugs. Fundam Clin Pharmacol 17:301–313

4. Wallace JL, McKnight W, Reuter B, Vergnolle N (2000) NSAID-induced gastric damage in rats: requirement for inhibition of both cyclooxygenase 1 and 2. Gastroenterology 119:705–714

5. Lichtenberger LM, Romero JJ, Dial EJ (2007) Surface phospholipids in gastric injury and protection when a selective cyclooxygenase-2 inhibitor (Coxib) is used in combination with aspirin. Br J Pharmacol 150:913–919

6. Whittle BJR (2004) Mechanisms underlying intestinal injury induced by anti-inflammatory COX inhibitors. Eur J Pharmacol 500:427–439

7. Masferrer JL, Zweifel BS, Manning PT, Hauser SD, Leahy KM, Smith WG, Isakson PC, Seibert K (1994) Selective inhibition of inducible cyclooxygenase 2 in vivo is anti-inflammatory and nonulcerogenic. Proc Natl Acad Sci U S A 91:3228–3232

8. Whittle BJR, Higgs GA, Eakins KE, Moncada S, Vane JR (1980) Selective inhibition of prostaglandin production in inflammatory exudates and gastric mucosa. Nature 284:271–273

9. Lanas A, Hunt R (2006) Prevention of anti-inflammatory drug-induced gastrointestinal damage: benefits and risks of therapeutic strategies. Ann Med 38:415–428

10. Laine L, Curtis SP, Cryer B, Kaur A, Cannon CP, MEDAL Steering Committee (2007) Assessment of upper gastrointestinal safety of etoricoxib and diclofenac in patients with osteoarthritis and rheumatoid arthritis in the Multinational Etoricoxib and Diclofenac Arthritis Long-term (MEDAL) programme: a randomized comparison. Lancet 369:465–473

11. Henry D, McGettigan P (2007) Selective COX-2 inhibitors: a promise unfulfilled? Gastroenterology 132:790–794

12. Funk CD, Fitzgerald GA (2007) COX-2 inhibitors and cardiovascular risk. J Cardiovasc Pharmacol 50:470–479

13. Whittle BJR (1977) Mechanisms underlying gastric mucosal damage induced by indomethacin and bile-salts, and the actions of prostaglandins. Br J Pharmacol 60:455–460

14. Konturek SJ, Brzozowski T, Piastucki I, Radecki T, Dembiński A, Dembińska-Kieć A (1982) Role of locally generated prostaglandins in adaptive gastric cytoprotection. Dig Dis Sci 27:967–971

15. Rainsford KD (1986) Structural damage and changes in eicosanoid metabolites in the gastric mucosa of rats and pigs induced by anti-inflammatory drugs of varying ulcerogenicity. Int J Tissue React 8:1–14

16. Gyömber E, Vattay P, Szabo S, Rainsford KD (1996) Role of early vascular damage in the pathogenesis of gastric haemorrhagic mucosal lesions induced by indomethacin in rats. Int J Exp Pathol 77:1–6

17. Gretzer B, Maricic N, Respondek M, Schuligoi R, Peskar BM (2001) Effects of specific inhibition of cyclo-oxygenase-1 and cyclo-oxygenase-2 in the rat stomach with normal mucosa and after acid challenge. Br J Pharmacol 132:1565–1573

18. Santos CL, Souza MH, Gomes AS, Lemos HP, Santos AA, Cunha FQ, Wallace JL (2005) Sildenafil prevents indomethacin-induced gastropathy in rats: role of leukocyte adherence and gastric blood flow. Br J Pharmacol 146:481–486

19. Satoh H, Inada I, Hirata T, Maki Y (1981) Indomethacin produces gastric antral ulcers in the refed rat. Gastroenterology 81:719–725

20. Satoh H, Ligumsky M, Guth PH (1984) Stimulation of gastric prostaglandin synthesis by refeeding in the rat. Role in protection of gastric mucosa from damage. Dig Dis Sci 29:330–335

21. Robert A (1975) An intestinal disease produced experimentally by a prostaglandin deficiency. Gastroenterology 69:1045–1047

22. Whittle BJR (1981) Temporal relationship between cyclooxygenase inhibition, as measured by prostacyclin biosynthesis, and the gastrointestinal damage induced by indomethacin in the rat. Gastroenterology 80:94–98

23. Reuter BK, Davies NM, Wallace JL (1997) Nonsteroidal anti-inflammatory drug enteropathy in rats: role of permeability, bacteria and enteropathic circulation. Gastroenterology 112:109–117

24. Evans SM, Whittle BJR (2001) Interactive roles of superoxide and inducible nitric oxide synthase in rat intestinal injury provoked by non-steroidal anti-inflammatory drugs. Eur J Pharmacol 429:287–296

25. Yokota A, Taniguchi M, Takahira Y, Tanaka A, Takeuchi K (2005) Rofecoxib produces intestinal but not gastric damage in the presence of a low dose of indomethacin in rats. J Pharmacol Exp Ther 314:302–309

26. Evans SM, Lazlo F, Whittle BJR (2000) Site-specific lesion formation, inflammation and inducible nitric oxide synthase expression by indomethacin in the rat intestine. Eur J Pharmacol 388:281–285

27. Evans SM, Whittle BJR (2003) Role of bacteria and inducible nitric oxide synthase activity in the systemic inflammatory microvascular response provoked by indomethacin in the rat. Eur J Pharmacol 461:63–71

28. Wallace JL, Whittle BJR (1986) Role of mucus in the repair of gastric epithelial damage in the rat. Inhibition of epithelial recovery by mucolytic agents. Gastroenterology 91:603–611

29. Skeljo MV, Giraud AS, Yeomans ND (1992) Adaptation of rat gastric mucosa to repeated doses of non-salicylate non-steroidal anti-inflammatory drugs. J Gastroenterol Hepatol 7:586–590

30. Clayton NM, Oakley I, Williams LV, Trevethick MA (1996) The role of acid in the pathogenesis of indomethacin-induced gastric antral ulcers in the rat. Aliment Pharmacol Ther 10:339–345

Chapter 17

Cyclooxygenase Enzymes and Their Products in the Carrageenan-Induced Pleurisy in Rats

Adrian R. Moore, Samir S. Ayoub, and Michael P. Seed

Abstract

Rodent models of inflammation have helped in our understanding of the inflammatory process and also for the screening of compounds with anti-inflammatory potential. Although they do not represent a particular inflammatory disease in humans, cavity models of inflammation in rodents are easy to induce and to quantify the inflammatory reaction as well as to harvest the inflammatory exudates for cytological, biochemical and molecular biological analysis. Of these models, the carrageenan-induced pleurisy model has been extensively used to study the role of the cyclooxygenase (COX) enzymes and the prostaglandins in acute inflammation and also for the screening of COX-inhibiting anti-inflammatory drugs.

Key words: Carrageenan, Inflammation, Pleurisy

1. Introduction

The inflammatory response is complex. It involves selective and temporal migration of specific cell types to the affected site, fluid exudation and usually (given time) some form of resolution. These events are orchestrated by a plethora of mediators that act sequentially to give rise to the ever-changing pattern of pathology. Many of these mediators are now identified and we are beginning to understand at a molecular level the intracellular signal transduction pathways activated when cells are exposed to these mediators.

The immune system of mammals is broadly similar across species. Inflammatory mediators in rodents usually have counterparts of high homology in man. Inflammatory models in rodents are therefore invaluable in investigating inflammatory mechanisms and for testing novel anti-inflammatory therapies that may be applicable to man. Nevertheless, care should always be exercised when

Samir S. Ayoub et al. (eds.), *Cyclooxygenases: Methods and Protocols*, Methods in Molecular Biology, vol. 644,
DOI 10.1007/978-1-59745-364-6_17, © Springer Science+Business Media, LLC 2010

extrapolating from animal models to man and certain mediators will vary enormously in their significance across species.

Rodent models of inflammation need to be robust and refined so as to allow inflammatory exudates to be harvested at various time points for cytological, biochemical and molecular biological analysis. In this regard, cavity models of inflammation offer distinct advantages over the ever-popular paw oedema (1) and cutaneous models (2). Inflammatory exudates can be harvested relatively simply from the peritoneum (3), pleural cavity (4) and air pouch (formed on the dorsum of an animal by the simple subcutaneous injection of air) (5). Inflammatory irritants can be directly injected into a cavity to produce a "non-immune" inflammatory response.

The choice of which cavity model of inflammation is suitable for a given study depends in part on the species used and the question to be addressed. Pleurisy is technically difficult in the mouse (although not impossible). Murine peritonitis is therefore a popular model. In rats, where pleurisy is straightforward, accurate quantification of exudates volume is easier than in peritonitis, where visceral organs interfere with exudate collection. One should, however, be aware that this model is subject to variations depending on the type of carrageenan used, its batch, amount injected, strain of animal, etc. As the carrageenan-induced pleurisy has been used extensively to study the roles played by the COX enzymes and the prostaglandins in inflammation, we decided to describe this model in this chapter.

Before we describe the model, we would like to briefly introduce some of the results that have been published on the expression profiles for COX-1 and COX-2 and on the prostaglandins in this model. The inflammatory reaction, measured as exudate formation and cell migration, peaks 24 h post-carrageenan injection into the pleural cavity. During the initial phase (0–12 h), the polymorphonuclear leucocytes dominate the cavity, which are then replaced by migrating mononuclear leucocytes that differentiate into macrophages. The carrageenan-induced pleurisy is a self-resolving inflammatory reaction that resolves after 48 h. The expression of COX-1 protein has been shown to remain unchanged during the entire 48 h time-course of the pleural inflammatory reaction. However, two peaks of increased expression of COX-2 have been shown to take place: one at the 2 h time-point and the second at the 48 h time-point. The levels of prostaglandin E_2 (PGE_2) were shown to coincide with the increase in the expression of COX-2 protein at the 2 h time-point. PGE_2 levels were shown to decrease thereafter and to remain low during the entire remaining course of inflammation. The expression of COX-2 during the 48 h time-point was shown to correlate with an increased release in prostaglandin D_2 (PGD_2) and its breakdown product 15deoxyΔ^{12-14}prostaglandin J_2 (15dPGJ$_2$). Inhibition of COX-2 activity during the early phase resulted in an anti-inflammatory effect, whereas inhibition of COX-2 activity during the resolving phase resulted in exacerbation of inflammation, suggesting that COX-2 during the resolving phase is associated with ant-inflammatory pathways (6).

2. Materials

All reagents may be purchased from Sigma Chemical Co, Poole, Dorset, UK, unless otherwise stated.

3. Methods

3.1. Carrageenan-Induced Pleurisy

1. Prepare a 1% solution of Lambda carrageenan in sterile saline (see Note 1).
2. Male Wistar rats (B&K Universal Ltd, Aldborough, Hull, U.K.) from about 180–220 g are ideal subjects. In any individual experiment, the weight range should be no more than about 20 g.
3. Anaesthetise the rats with a suitable gaseous anaesthetic such as halothane or isofluorane.
4. The animal is laid on its right flank and the skin over the thorax moistened with 70% ethanol.
5. A small cut is made to expose the intercostals muscles.
6. A 21 gauge needle that has been reduced in length to about 6 mm (bend and break the needle at the appropriate point) is used to inject 0.15 ml of 1% carrageenan into the pleural cavity at the level of the fifth to sixth intercostals space (see Note 2).
7. Close the wound with a wound closure clip or with tissue glue and allow the animals to recover from the anaesthetic.
8. At time intervals appropriate to the experiment, kill the animals by over-exposure to anaesthetic.
9. The musculature of the thorax and upper abdomen is exposed and the muscle incised to reveal the xiphisternum. This should be gripped firmly with forceps and the diaphragm perforated just below the cartilage. Two cuts either side of the sternum enable a flap to be reflected to reveal the thoracic cavity.
10. 1 ml of phosphate buffered saline (PBS) containing an anticoagulant such as heparin at 10 U/ml is introduced to the cavity (see Note 3).
11. Using a plastic Pasteur pipette, the exudates can be recovered into suitable tubes.
12. The volume of recovered exudates can be determined either by collection into graduated tubes or by collecting into weight matched tubes and noting the change in weight (see Notes 4–6).

13. Cells may be counted by hand using a hemocytometer or more conveniently using a Coulter counter.

14. The recovered exudate volume multiplied by the cell count/ml gives the total cells.

15. The true exudate volume can be calculated from the recovered exudate volume by subtraction of the 1 ml lavage volume.

16. Differential cell counts can be performed on smears or cytospin preparations (see Notes 7–9).

3.2. Administration of COX Inhibitors

1. The following COX inhibitors have been investigated in this model (6 and unpublished data); NS398 (1–10 mg/kg), celecoxib (1–15 mg/kg), piroxicam (0.1–10 mg/kg), nimesulide (0.5–5 mg/kg), indomethacin (0.3–3 mg/kg), aspirin (10–200 mg/kg), SC560 (1–15 mg/kg) and paracetamol (50–200 mg/kg; see Notes 10 and 11).

2. All drugs are prepared in 1% gum tragacanth dissolved in distilled water (see Note 12).

3. Drugs are administered by the oral route.

4. Drugs should be administered 1 h before carrageenan in order to study their actions during the early phase of inflammation and at the 24 h time-point (and again every 6 h) to investigate their effect on the 48 h resolution phase (see Note 13).

4. Notes

1. Carrageenan is a natural extract from seaweed and will be subject to batch variation. It exists in many forms; lambda carrageenan is described as non-gelling. Carrageenan provokes a complement-dependent inflammation.

2. The blunted needle described above creates a resistance when pushing through into the pleural space. A distinct "pop" is felt as the needle passes through allowing the experimenter to know that the needle is correctly placed. The rounded tip also minimises the chances of injecting directly into the lungs.

3. Ex vivo production of inflammatory mediators into exudates may be minimised by incorporating appropriate inhibitors in the lavage solution. For example, $10\,\mu M$ indomethacin is often included in the lavage when exudates are to be assayed for cyclooxygenase products.

4. Blood contaminated exudates should be rejected.

5. For biochemical analysis, exudates should be collected onto ice and centrifugation used to prepare cell free exudate and cell pellets.

6. It may be necessary to include a standard protease inhibitory cocktail in the lavage solution to protect polypeptide mediators, enzymes and other proteins from degradation.

7. Cell pellets treated in this way are suitable for Western blotting.

8. In addition to routine cytological analysis, cell preparations may be used for immunocytochemistry, enzyme cytochemistry or in situ hybridisation.

9. Exudate cells may be removed for cell culture ex vivo. In which case, cell culture medium may be more appropriate than PBS for lavage.

10. Paracetamol produced no anti-inflammatory effects during the initial phase, but it exacerbated the inflammatory reaction when administered at 24 h and then every 6 h and inflammation measured at 48 h (our unpublished observations).

11. The inhibitors listed are a selection of NSAIDs with different inhibitory potencies for COX-1 and COX-2. COX-2 selective inhibitors (celecoxib and NS398), COX-2 preferential inhibitor (peroxicam), non-selective inhibitors (indomethacin and aspirin), COX-1 selective inhibitor (SC560). These compounds at the doses listed can be used as positive controls. SC560 does not affect the inflammatory reaction in any way.

12. Preparation of gum tragacanth involves continuous stirring with gentle heating (45°C) with gum tragacanth powder added slowly.

13. The 6-hourly dosing regime for drugs during the resolution phase is only a guide. Drugs with a long plasma half-life should be administered less frequently.

References

1. Winter CA, Risley EA, Nuss GW (1963) Anti-inflammatory and antipyretic activities of indomethacin 1-(p chlorobenzoyl)-5-methoxy-2-methyl-indole-3-acetic acid. J Pharmacol Exp Ther 141:369–376

2. Eipert EF, Miller HC (1975) Contact snsitivity in mice measured with thymidine labeled lymphocytes. Immunol Commun 4:361–372

3. Murch AR, Papadimitriou JM (1981) The kinetics of murine peritoneal macrophage replication. J Pathol 133:177–183

4. Spector WG, Willoughby DA (1957) Histamine and 5 hydroxytryptamine in experimental pleurisy. J Pathol Bacteriol 74:57–65

5. Selye H (1953) On the mechanism through which hydrocortisone affects the resistance of tissue to injury. J Am Med Assoc 152:1207–1213

6. Gilroy DW, Colville-Nash PR, Willis D, Chivers J, Paul-Clark MJ, Willoughby DA (1999) Inducible cyclooxygenase may have anti-inflammatory properties. Nat Med 5:698–701

Iloprost-Induced Nociception: Determination of the Site of Anti-nociceptive Action of Cyclooxygenase Inhibitors and the Involvement of Cyclooxygenase Products in Central Mechanisms of Nociception

Samir S. Ayoub and Regina M. Botting

Abstract

The writing response to acute nociception has been used to test the analgesic activity of drugs in rodents. Dilute acetic acid is the most frequently used irritant to induce writhing behaviour. The administration of acetic acid intraperitoneally activates both peripheral and central mechanisms of nociception. It releases nociceptive mediators such as prostaglandins (PG) E_2 and I_2 at the site of noxious stimulation, the peritoneal cavity, and at central sites such as the dorsal horn of the spinal cord and some brain regions. We have used the PGI_2 mimetic, iloprost, an agonist at the IP receptor, to induce the writhing response in mice. Iloprost activates the IP receptors on peripheral nociceptors directly and thus does not release nociceptive prostaglandins into the peritoneal cavity. However, prostaglandins are still involved in nociceptive transmission at the spinal and supraspinal levels. Using this model of nociception, it is possible to identify the site of action of analgesic drugs which reduce prostaglandin release in central tissues through inhibition of cyclooxygenase. Thus, a drug that inhibits the iloprost-induced writhing response and reduces release of prostaglandins in the central nervous system is likely to be a centrally acting analgesic drug. This chapter compares the iloprost- and acetic acid-induced writhing responses in mice and describes a method for measuring central prostaglandin levels. Part of this work has been published previously (1).

Key words: Iloprost, Writhing, Cyclooxygenase, Central nervous system, Prostaglandin E_2

1. Introduction

Animal models of nociception constitute a major part of pre-clinical studies in the search for new analgesic drugs. A large number of different animal models of nociception are available, each representing a different pain state in humans, such as inflammatory

Samir S. Ayoub et al. (eds.), *Cyclooxygenases: Methods and Protocols*, Methods in Molecular Biology, vol. 644, DOI 10.1007/978-1-59745-364-6_18, © Springer Science+Business Media, LLC 2010

and non-inflammatory pain, hyperalgesia, neuropathic pain and visceral pain (2).

The writing test to determine acute nociception in rodents was first described in rats by Vander and Margolin in 1956 (3). The writing response is induced by the intraperitoneal injection of a noxious substance such as dilute acetic acid. It consists of a wave of constriction and elongation passing caudally along the abdominal wall followed by extension of the hind limbs (4). The writing test is used in laboratory studies for the screening of analgesic drugs (5, 6) and for the study of pain mechanisms. Due to ethical and emotional constraints, this test has been referred to as the "abdominal constriction test." However, "writing" is still the preferred term together with "stretching" and "cramping." A large number of chemical irritants have been used to induce writing in rodents such as phenylbenzoquinone, acetyl choline, bradykinin and others (4). The writing behaviour in response to acetic acid has also been associated with loss of co-ordination (7) and reduced locomotion (8), both effects being reversed by analgesic drugs. As shown by Adachi in 1994 (9), the writing response disappears after about an hour following the injection of acetic acid. In addition, Doherty et al (1985) (10) showed that zymosan injection which leads to writing within minutes, results in inflammation, measured by extravasation and accumulation of plasma proteins and migration of neutrophils, 3 h after.

Analysis of this experimental model confirmed that the writing behaviour is truly due to a nociceptive response to the chemical irritant. Thus, various nociceptive mediators were shown to be involved, of which the prostaglandins have been demonstrated to play a major role. Gyires and Knoll (1975) (11) were the first to show this, by inducing writing with an injection of prostaglandin E_1 (PGE_1) and prostaglandin E_2 (PGE_2) and diminishing the response with drugs that reduce prostaglandin levels. The prostaglandins were also shown to be involved in initiating the response in other commonly used writing models induced by such chemicals as acetic acid (12), phenylbenzoquinone (13), zymosan (10, 14) and kaolin (15). Additionally, most of these authors reported a greater increase in prostaglandin I_2 (PGI_2, measured as its stable metabolite 6-keto-$PGF_{1\alpha}$) than PGE_2 in the peritoneal fluid after the induction of writing. More recently, it was demonstrated that writing responses cannot be induced in response to acetic acid in IP receptor knockout mice (16). Both PGE_2 and prostaglandin D_2 (PGD_2) administered intracisternally to animals writing to acetic acid produced a biphasic effect. Thus, at low doses (5–15 ng) these prostaglandins had a hyperalgesic effect and at high doses (>50 ng PGD_2 and 5 μg PGE_2) they demonstrated a hypoalgesic action (17).

2. Materials

2.1. Animals

Female, C57BL/6 mice weighing 20–25 g obtained from Harlac-Olan Ltd, UK were maintained under 12 h/12 h light/dark cycle at $22 \pm 1°C$. Food and water were provided ad libitum. The animals were acclimatised to the experimental room prior to testing.

2.2. General Reagents

1. Chromatographic purity glacial acetic acid is diluted to 0.6% (v/v) with physiological saline.

2. Celecoxib 4-[5-(4-methylphenyl)-3- (trifluoromethyl)-1H-pyrazol-1-yl] benzenesulfonamide is dissolved in dimethyl sulphoxide (DMSO) and diluted further with a solution consisting of 10% cremophor EL, 10% ethanol and 80% physiological saline bringing the final concentration of DMSO to below 1%.

3. C18 Sep-Pak columns (Waters Inc., U.S.A.).

4. Diclofenac sodium (2-[((2,6-dichlorophenyl)amino)]benzeneacetate sodium) is dissolved in 0.9% saline.

5. Ethyl acetate: (chromatographic purity), 99.9% for HPLC.

6. Iloprost (5-[Hexahydro-5-hydroxy-4-(3-hydroxy-4-methyl-1-octen-6-ynyl)-2(1H)-pentalenylidene) pentanoic acid) was obtained in ampoules of 0.5 mg iloprost/ml and diluted to 2.5 µg/ml with physiological saline. These ampoules were a gift from Schering AG. Berlin, Germany.

7. Paracetamol (acetaminophen, 4-amidophenol) is dissolved in 12.5% (v/v) 1,2-propanediol.

8. SC560 (5-(-4-chlorophenyl)-1-(4-methyloxyohenyl)-3-(trifluoromethyl)-1H-pyrazole) is dissolved in DMSO and diluted further with a solution consisting of 10% cremophor EL, 10% ethanol and 80% physiological saline bringing the final concentration of DMSO to below 1%.

9. 15% (v/v) ethanol is acidified to pH 3 with concentrated HCl.

2.3. Prostaglandin E$_2$ Enzyme Immunoassay (EIA) (from GE Healthcare)

1. Assay buffer is diluted in 500 ml of distilled water.

2. Wash buffer is diluted in 500 ml of distilled water.

3. Lyophilised anti-PGE$_2$ antibody is initially diluted in 6 ml assay buffer and then 2 ml of diluted antibody is added to 4 ml assay buffer.

4. Lyophilised PGE$_2$ conjugate is initially diluted in 6 ml assay buffer and then 2 ml of diluted conjugate is added to 4 ml assay buffer.

5. PGE$_2$ standard was serially diluted as follows: 100 µl of stock standard (256 ng/ml) is diluted in 3.9 ml of assay buffer to give the top standard with a concentration of 320 pg/50 µl.

Seven tubes are labelled 2.5, 5, 10, 20, 40, 80, 160 pg/50 µl and 500 µl of assay buffer is added to each. To the 160 pg/50 µl tube, 500 µl of 320 pg/50 µl PGE$_2$ is added and mixed thoroughly. To the 80 pg/50 µl tube, 500 µl of 160 pg/50µl is added and mixed thoroughly. This doubling dilution is repeated with the remaining tubes.

3. Methods

3.1. Induction of Acetic Acid or Iloprost-Induced Writhing

1. C57Bl/6 mice weighing 20–25 g are placed in round, open topped, clear perspex cylinders, 30 cm in diameter and with walls 15.5 cm high 5 min before iloprost or acetic acid injection.

2. Induce the writhing behaviour in mice by the intraperitoneal injection of 0.1 ml/10 g body weight of either 0.6% (v/v) glacial acetic acid or 0.25 µg/10 g body weight of iloprost (See Note 1).

3. Count the number of writhes, abdominal extensions or abdominal constrictions for a 20 min period following the administration of the noxious substance (See Note 2).

3.2. Drug Administration

1. For the assessment of analgesic activity, administer drugs by the subcutaneous (See Note 3) route prior to the induction of writhing with either acetic acid or iloprost.

2. For optimal antinociceptive effect, 40–400 mg/kg paracetamol (Figs. 1 and 2) or 5–100 mg/kg diclofenac sodium are administered 30 min prior to induction of writhing.

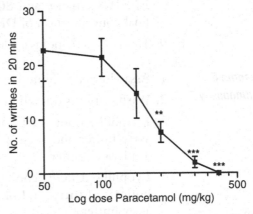

Fig. 1. This figure illustrates the effect of paracetamol on acetic acid-induced writhing in C57Bl/6 mice. The number of writhing responses induced in 20 min after an intraperitoneal injection of 0.6% acetic acid was reduced dose-dependently by paracetamol (100–400 mg/kg) injected subcutaneously 30 min before. Each point represents the mean ± S.E.M.; ** = $P < 0.01$; *** = $P < 0.001$ compared to vehicle. $N = 6$–10 animals in each group.

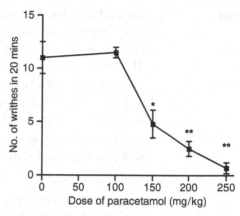

Fig. 2. This figure illustrates the effect of paracetamol on iloprost-induced writhing in C57Bl/6mice. The number of writhing responses induced in 20 min after an intraperitoneal injection of 0.25 μg iloprost per 10 g body weight in 0.1 ml physiological saline was reduced by paracetamol (100–250 mg/kg) injected subcutaneously 30 min before. Each point represents the mean ± S.E.M.; *=$P<0.05$; **=$P<0.01$ compared to vehicle. N = 6–10 animals in each group.

The COX-1 and COX-2 selective inhibitors SC560 (1–15 mg/kg) or celecoxib (1–15 mg/kg) are administered 60 min before injection of acetic acid or iloprost (See Note 4: how pre-dosing was determined).

3.3. Tissue Collection

1. At the end of the 20 min observation period, the animals are killed by cervical dislocation.

2. When writhing had been induced with acetic acid, peritoneal fluid is collected for measurement of PGI_2 and PGE_2 concentrations. Wash the peritoneal cavity by injecting 1 ml of 0.9% (w/v) saline. Expose the cavity by a longitudinal incision of the peritoneal wall and collect the fluid with a syringe. Reduction in PGI_2 and PGE_2 levels provides an assessment of possible inhibition of peripheral COX activity.

3. To remove the spinal cord, make an incision on the dorsal and cranial ends of the spinal column. Eject the complete spinal cord from the spinal column using a saline-filled 1 ml syringe by inserting a 23G needle into through the dorsal side of the spinal column. Snap-freeze the spinal cord in liquid nitrogen.

4. Remove the whole brain from the skull with a spatula or dissect anatomically into the cerebral cortex, mid brain, brain stem and cerebellum and snap-freeze in liquid nitrogen.

5. Keep the peritoneal fluid on ice and then spin down at 350 × g for 10 min at 4°C and store the supernatant at −80°C.

6. Crush spinal cord and brain tissues to powder using a nitrogen bomb (supplied by Biospec Products, Bartlesville, OK, U.S.A.) previously cooled in liquid nitrogen. Store powdered tissues at −80°C.

3.4. Prostaglandin Extraction

1. Suspend powdered spinal cord or brain tissues in 1 ml of 15% (v/v) ethanol (adjusted to pH 3 with concentrated HCl), leave on ice for 10 min and then spin down at $375 \times g$ for 10 min at 4°C.

2. Add 500 μl of 15% (v/v) ethanol (at pH 3) to 500 μl peritoneal fluid, leave on ice for 10 min and then spin down at $375 \times g$ for 10 min at 4°C.

3. Condition C-18 Sep-Pak columns with 4 ml ethanol followed by 4 ml of distilled water at a flow rate of 5–10 ml/min.

4. Apply samples to the columns at a flow rate of 5 ml/min.

5. Wash the columns with 4 ml of distilled water followed by 4 ml of 15% (v/v) ethanol (at pH 3).

6. Elude the samples with 2 ml of ethyl acetate at a flow rate of 5 ml/min.

7. Dry the samples in a spin vaccum.

3.5. Prostaglandin E₂ Measurements

1. Set-up a 96 well plate coated with goat anti-mouse IgG with sufficient wells to run the standards, controls and samples.

2. Add to the non-specific binding (NSB) wells 100 μl of assay buffer.

3. Add to the zero standard wells 50 μl of assay buffer.

4. Starting with the most dilute sample, add 50 μl of each standard PGE₂ in duplicate.

5. Add 50 μl of samples diluted appropriately to the plate in duplicate.

6. Add 50 μl of diluted antibody to all the wells except the blank (containing nothing) and NSB.

7. Add 50 μl of diluted conjugate to all the wells except the blank.

8. Cover the plate and incubate at room temperature for 1 h on a microplate shaker.

9. At the end of the 1h incubation period, wash the plate four times with wash buffer by filling and completely empting the plate between washes.

10. Add 150 μl of substrate to all wells on the plate and incubate the plate at room temperature for 30 min.

11. Read the blue colour developed at 630 nm absorbance.

3.6. Calculations of Results

1. Initially calculate the percentage binding ($\%B/B_0$) for standards and samples using the following equation:

$$\%B \, / \, B_0 \, = \, \frac{(standard/sample \, OD \, - \, NSB \, OD)}{(B_0 OD \, - \, NSB \, OD)} \times 100$$

Where: B_0 = reading in zero standard wells.

OD = Optical density (absorbance values at 630 nm)

NSB = Non-specific binding wells.

2. Generate a standard curve by plotting the $\%B/B_0$ as a function of the log concentration of PGE_2.

3. Read the $\%B/B_0$ for the samples directly from the standard curve.

4. Measurements of PGE_2 levels in spinal cord (Fig. 3) and whole brain (Fig. 4) demonstrated a dose related reduction in PGE_2 levels following paracetamol administration (Figs. 3, 4).

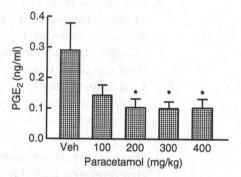

Fig. 3. This figure illustrates the prostaglandin E_2 concentrations in the spinal cord of mice injected with paracetamol. Concentrations of prostaglandin E_2 (PGE_2) in extracts of spinal cord were reduced in mice injected subcutaneously with increasing doses of paracetamol (100–400 mg/kg) compared to mice injected with vehicle only. Histograms represent mean ± S.E.M. *Veh* vehicle; * = $P < 0.05$; N = 6–10 animals in each group.

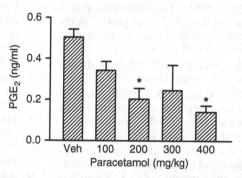

Fig. 4. This figure illustrates the prostaglandin E_2 concentrations in the brains of mice injected with paracetamol. Concentrations of prostaglandin E_2 (PGE_2) in extracts of brain were reduced in mice injected subcutaneously with increasing doses of paracetamol (100–400 mg/kg) compared to mice injected with vehicle only. Histograms represent mean ± S.E.M. *Veh* vehicle; * = $P < 0.05$; N = 6–10 animals in each group.

4. Notes

1. After the intraperitoneal administration of acetic acid, there is a delay of approximately 5 min before the mice begin to exhibit writing behaviour. After administration of iloprost, the delay is approximately 2 min.

2. Occasionally after injection of acetic acid, but more often after administration of iloprost, mice sometimes exhibit abdominal elongations without extension of the hind limbs. We describe abdominal elongations with extension of the hind limbs as full writhes and abdominal elongations without extension of the hind limbs as half writhes. It is important for the observer to standardise the way writing responses are counted. Thus, half writhes may be excluded from the count or two half writhes can be counted as one full writhe.

3. It is best to avoid intraperitoneal administration of test drugs as this is the same site as the administration of the noxious substance and can interfere with the writing count. For example, the vehicle used to dissolve the drug or the drug itself can cause more irritation to the already sensitised area. Oral administration of paracetamol has been shown to dramatically increase the writing counts. Thus, we found that subcutaneous administration of test drugs was the most satisfactory route to employ.

4. Previous experimentation with different pre-dosing intervals demonstrated that the optimal pre-dosing period for maximum analgesic activity of test drugs was 30 min.

References

1. Ayoub SS, Colville-Nash PR, Willoughby DA, Botting RM (2006) The involvement of a cyclooxygenase-1 gene-derived protein in the antinociceptive action of paracetamol in mice. Eur J Pharmacol 538:57–65

2. Stanley KL, Paice JA (1997) Animal models in pain research. Semin Oncol Nurs 13:3–9

3. Vander WC, Margolin S (1956) Analgesic tests based upon experimentally induced acute abdominal pain in rats. Fedn Proc 15:494

4. Collier HOJ, Dinneen LC, Johnson CA, Schneider C (1968) The abdominal constriction response and its suppression by analgesic drugs in the mouse. Br J Pharmacol Chemother 32:295–310

5. Siegmund E, Cadmus R, Lu G (1957) A method for evaluating both non-narcotic and narcotic analgesics. Proc Soc Exp Biol Med 95:729–731

6. Sewell RDE, Gonzalez JP, Pugh J (1984) Comparison of the relative effects of aspirin, mefenamic acid, dihydrocodeine, dextropropoxyphene and paracetamol on visceral pain, respiratory rate and prostaglandin biosynthesis. Arch Int Pharmacodyn 268:325–334

7. Singh PP, Junnarkar AY, Varma RK (1987) A test for analgesics: incoordination in writing mice. Meth & Find Exptl Clin Pharmacol 9:9–11

8. Harada T, Takahashi H, Kaya H, Inoki R (1979) A test for analgesics as an indicator of locomotor activity in writing mice. Arch Int Pharmacodyn 242:273–284

9. Adachi KI (1994) A device for automatic measurement of writing and its application to the assessment of analgesic agents. J Pharmacol Toxicol Meth 32:79–84

10. Doherty NS, Poubelle P, Borgeat P, Beaver TH, Westrich GL, Schrader NL (1985)

Intraperitoneal injection of zymosan in mice induces pain, inflammation and the synthesis of peptidoleukotrienes and prostaglandin E_2. Prostaglandins 30:769–789

11. Gyires K, Knoll J (1975) Inflammation and writhing syndrome inducing effect of PGE_1, PGE_2 and the inhibition of these actions. Pol J Pharmacol Pharm 27:257–264

12. Drower EJ, Stapelfeld A, Mueller RA, Hammond DL (1987) The antinociceptive effects of prostaglandin antagonists in the rat. Eur J Pharmacol 133:249–256

13. Berkenkopf JW, Weichman BM (1988) Production of prostacyclin in mice following intraperitoneal injection of acetic acid, phenylbenzoquinone and zymosan: its role in the writhing response. Prostaglandins 36:693–709

14. Doherty NS, Beaver TH, Chan KY, Coutant JE, Westrich GL (1987) The role of prostaglandins in the nociceptive response induced by intraperitoneal injection of zymosan in mice. Br J Pharmacol 91:39–47

15. Fujiyoshi T, Kuwashima M, Iida H, Uematsu T (1989) A new writhing model of factor XII activator-induced pain for assessment of non-steroidal anti-inflammatory agents. I. Kaolin-induced writhing in mice. J Pharmacobio-Dyn 12:132–136

16. Murata T, Ushikubi F, Matsuoka T, Hirata M, Yamasaki A, Sugimoto Y, Ichikawa A, Aze Y, Tanaka T, Yoshida N, Ueno A, Oh-ishi S, Narumiya S (1997) Altered pain perception and inflammatory response in mice lacking prostacyclin receptor. Nature 388:678–682

17. Horiguchi S, Ueno R, Hyodo M, Hayaishi O (1986) Alterations in nociception after intracisternal administration of prostaglandin D_2, E_2, or $F_{2\alpha}$ to conscious mice. Eur J Pharmacol 122:173–179

association of asparagine with the stability of
amino phosphatase and glutathione to prevent undesirable change and alternate the
effect of acidity present in the —

11. Marshall K. Rose, H. C. Johnson and Haber, C. (1961) The influence and role of
lead, and metabolism of — "
Biochem. J. Physiol. 110."

12. Johnson, D., Slagle, K. B., Thomas, A.,
Edmund, D. K. (1987) The association of —
effect of morphological association in the rat
Int. J. Physiol. Biochem. 58.

13. Gallagher, N. H., Newberger, E. D. (1986)
Functions of association in mice, including
management, nutrition and storage. Including glycerol, proteins and lipoprotein in rat
int. Biochem. Soc.

14. Jackson, W. J. and H. Smith J. (Juchter in
Winchell O. G. (1937) The evaluation of morphogenesis

On the morphology of Chondrus in cells accompanied by experimental hypothesis of new transporto-
acidity of — "Biochem J." 58.

15. Marshall, J., Sakata and H., Hull, J. Johnson
D. H. Rector (1988) Association of the effect of the —
new association and transport of nutrients as Nutrition
of nutrients as transport as protein — Thomas, John, Pwn
J. Biochem. 110."

16. Michel, A. K., R. H. C., Johnson and T. Rector, M.
(1988) Association, transport, stability. Johnson, A. Rector,
Pwn, E. A., Slagle, K. Longer, N. C. Johnson.
J. Biochem 58, and J. 58. Pwn. Association and transport
of nutrients association as the transport in mice including
morphology, nutrients, Johnson 58. Bioch. 58.

17. Richardson, M. E. and Rector, M. H. Johnson, O.
(1986) Association in storage of new protein
acid administration of association in 25,
J. Biochem to transport as J. Phys. Johnson Or.
Phys. Biochem 110."

INDEX

Printed in the United States
By Bookmasters